I AM NOW BACK ACROSS THE 38TH PARALLEL LINE

Letters from a 5th Regimental Combat Team Soldier

JULIE MULLENAX VAN METER

INDIE APPALACHIA LLC
NORTH MANKATO, MN

I am now back across the 38th Parallel Line
Letters from a 5th Regimental Combat Team Soldier
First edition: November 2025
© 2025 Julie Mullenax Van Meter

All Rights Reserved.
This book is subject to the condition that no part of this book is to be reproduced, transmitted in any form or means; electronic or mechanical, stored in a retrieval system, photocopied, recorded, scanned, or otherwise. Any of these actions require the proper written permission of the author.

Cover Photo: Dustin Eye
Book Design: Mike Mallow

Printed in the U.S.A.
ISBN: 979-8-9926451-3-2 (Paperback)
ISBN: 979-8-9926451-4-9 (Hardback)
ISBN: 979-8-9926451-5-6 (Digital)
Printed by Indie Appalachia LLC

CONTENTS

Chapter One: The Beginning 9
Family Genealogy 10
Chapter Two: WW II: The Wars of all Wars 11
Medals Awarded 13
Chapter Three: The Korean Conflict 15
Medals Awarded 17
Chapter Four: Reservists are Called for Training 19
Letter 1 to Letter 88 20
Chapter Five: Leaving Ft. Campbell KY. to Camp Stoneman CA 129
Letter 89 to Letter 100 130
Chapter Six: From Camp Stoneman CA. to the Pursan Perimeter, South Korea 149
Letter 101 to Letter 169 150
Chapter Seven: Orders to Ship Back to the US 245
Letter 170 to Telegram 175 246
Chapter Eight: Home – Life After Korea 253
Other Sources 255

Sgt. John W. Mullenax

DEDICATION

I would like to dedicate this book to all Veterans past and present. The patriotism you have demonstrates the courage and importance of keeping the world safe.

TO MY SON

Words cannot express the love, respect and devotion I have for you. You are wise beyond your years. I am so proud. You have accomplished so much, and I hope you keep climbing that ladder to the top of your world. If it weren't for you, this book would not have been completed. You gave me the desire to write it for everyone to read and have a visual of what happened during the Korean Conflict. I love you John Dustin Eye.

THANK YOU

To my loving husband Rodney Van Meter, cousins, Barbara Bennett, Janet Mullenax, and Michael Mallow and too many to mention. Thank You so much.

A SPECIAL THANK YOU

I would like to thank Sam Kier, Historian of the 5th RCT for his help, patience and guidance. Without his help, this book would not have been completed. He wrote "Two Wars, Two Bronze Stars, Two Purple Hearts" about my dad in the Fall 2022 Bobcat Bulletin.

John W. Mullenax (left) and Henry Walter Mullenax (father) pose in early summer 1953 with what is considered the widest log ever taken out of the woods. The log was displayed in the local parade on a 1954 Chevrolet to advertise the trucks being offered for sale that fall. (Courtesy of Janet Mullenax)

CHAPTER ONE
THE BEGINNING

Located in Virginia, hence lies the County of Highland. It is one of the 25 Counties in Virginia that is part of the Appalachian Region. It has been named "Virginia's Little Switzerland." Highland was formed from portions of the counties of Bath and Pendleton. It was passed by the Virginia General Assembly on March 19, 1847.

The small village of Crabbottom was built. Crabbottom consisted of three General Stores, a water mill, two churches, a high school, several shops, and about 20 families.

A county-run "Poor Farm" from the 1800s provided shelter, clothing, food, and medicine for the poor and indigent. This farm was run by three generations of Mullenaxes for close to 100 years. The Poor Farm was self-sufficient. It was located on 135 acres, providing space for wheat, barley, oats, vegetables, cattle, sheep, chickens, and hogs. The indigents would work the farm for their room and board.

John William Mullenax was born on February 12, 1922, at the Poor Farm in Crabbottom, Virginia. He was the youngest child of Henry Walter Mullenax and Mamie Catherine Collins Mullenax. He had 10 Siblings, 3 brothers, and 7 sisters.

He attended Crabbottom School and stayed in school until after 10th grade. Like his brothers, John worked the Farm. As he grew older, he would learn how to play the guitar and banjo.

On Saturday nights, Henry Walter, Joe and John would go to dances and play. Henry Walter would play the fiddle, Joe the guitar, and John either the guitar or banjo. He also hunted deer and bears and went fishing in the many brooks in the area.

After he returned home from WWII, John became his father's only business partner in the sawmill, Mullenax Lumber Company, by buying out his brothers' interests.

One evening in May 1950, John met a young lady at the local Loyal of Order Moose in Durbin, West Virginia. Edith Mozella Parsons Zinkhan was living with her mother and stepfather. Edith was previously married to Benjamin C. Zinkhan, Jr., who died in WWII, leaving her with a young son Benjamin III to raise.

FAMILY GENEALOGY

Born: John William Mullenax – February 12, 1922 / December 6, 1994

He is the 11th Child & 4th Male from the Parents of:

Father: Henry Walter Mullenax & Mother: Mamie Katherine Collins Mullenax

1. Esta Belle Mullenax
2. Monna Lee Mullenax Hammer
3. Fred B Mullenax
4. Hazel Gray Mullenax
5. Gladys May Mullenax Swecker
6. Thelma Gertrude Mullenax Mullenax
7. Edith Virginia Mullenax
8. Geneva Katherine Mullenax Fox
9. Joseph Thayer Mullenax
10. Benjamin Walter Mullenax
11. John William Mullenax

CHAPTER TWO

WWII: THE WARS OF ALL WARS

On April 13, 1944, John traveled to Roanoke, VA, and registered with the Selective Service, just like his brothers before him.

He was inducted at Ft. George G. Meade in Maryland. There he trained and entered Company F, 334th Infantry Regiment of the 84th infantry Division, as a private. He sailed to Scotland on October 5, 1944, and arrived there on October 13, 1944. On October 31, 1944, he left Scotland and traveled to France. November 1, 1944, he went ashore at Omaha Beach in Normandy. November 7, 1944, he traveled through France in a truck convoy. November 8, 1944, He could hear artillery firing on Germany's western boundary, the Siegfried Line. November 18, 1944, the 84th Division crosses the Siegfried Line into Germany and captures the city of Geilenkirchen.

On November 30, 1944, the 334th Infantry Regiment, 84th Infantry Division, received orders to seize an area of high ground north of Aachen, Germany. It was the Unit's 12th day in combat. The men of the Second Battalion came under artillery and mortar fire long before they reached their line of departure.

Fox Company was ordered to attack the German defenses near the little town of Apweiler at 7:30 on November 30, 1944. By then, it had taken heavy casualties and had become somewhat disorganized. They ran into a German artillery barrage that stalled the attack.

Twenty-two-year-old Private John W. Mullenax from Blue Grass, VA, was wounded at 8 00 p.m. by shrapnel that penetrated the left wrist. He was shipped to the 105th Evacuation Hospital in Maastricht, Holland, where he underwent surgery. He was transferred to the 28th General Hospital in Dorsetshire, England, for a few weeks of rehabilitation.

Mamie Mullenax received a letter from the War Department stating that her son had been slightly wounded in action in Germany.

Mullenax returned to the 334th on February 2, 1945, the assembly area in Holland, getting their first rest since they entered a combat zone on November 18, 1944. While receiving rehabilitation, the 84th Division "Rail Splitters" had helped stem the German counter-attack in the Ardennes region of Belgium (Battle of the Bulge). It was time to renew the attack and chase the German Army back to Germany.

The 84th Division crossed the Rohr River on February 23, 1945, and before crossing the Rhine, was ordered to occupy Homberg on the west bank of the Rhine. The citizens of that ancient German City wanted the obviously skeptical G.I.s to know that there were "no Nazis in Homberg. All the folks in town have hated Hitler for a long time, and they all have cousins in Chicago."

On April 1, Second Battalion secured the bridges into the city of Hannover, and the regiment crossed the Rhine River. They soon found the Hannover-Ahlem prison camp and released and arranged for the care of several starving and ill Jewish prisoners. The healthier inmates had been removed from the camp by the German SS prior to the arrival of the 334th Infantry and had marched off to an extermination site.

The 334th continued East until they reached the Elbe River on April 14. They were ordered to remain there until the Russian Army arrived on the East Bank of the Elbe. That occurred on May 2. The 84th Division's campaign in Europe had come to an end. The war ended in Europe six days later.

They remain in Germany on occupation duty until January 1946. They began coming home in January 1946. Thirty-four members of Company F had given their lives to free Europe from German occupation.

Because they had not been part of the June 6, 1944 invasion farce, few members of the 84th Infantry Division had sufficient rotation points to return to the U.S. as soon as peace in Europe was declared.

The Division remained in Germany to serve as a peacekeeping force.

In May 1946, John sailed home and reported to the Separation Center at Ft. Meade, KY. He was discharged from the Army Rank of staff sergeant. At that time, he joined the Army Reserves. He was able to retain his active-duty rank of staff sergeant.

World War II was a global military conflict that was fought between September 1, 1939, and September 2, 1945. The War has pitted two major military alliances against each other: the Allies of the U.S., Soviet Union, United Kingdom, and China and others against the Axis of Germany, Japan, Italy, and others.

Over 60 million people, the majority of them civilians, were killed, making it the deadliest conflict in human history. WWII was known for modern warfare and tactics such as air warfare, strategic bombing, and the first and only use of nuclear weapons in warfare.

Medals Awarded

1. Purple Heart
2. Bronze Star
3. European African Middle Eastern Theater Ribbon
4. Good Conduct Medal
5. WWII Victory Ribbon
6. Army Occupation Medal
7. M1 Rifle Combo Infantry Badge
8. Rifle Expert Medal

CHAPTER THREE

THE KOREAN CONFLICT

In 1948 the occupation zones became two sovereign states. North Korea was a Socialist state named the Democratic People's Republic of Korea. Under the Communist leader of Kim Il-sung.

South Korea was a Capitalist state; the Republic of Korea was established in the south under the authoritarian leadership of Syngman Rhee.

The United Nations Security Council denounced North Korea's move as an invasion, authorized the formation of the United Nations Command, and dispatched the forces of Korea to repel it.

The Soviet Union was boycotting the U.N. for recognizing Taiwan (Republic of China) as China and China (Peoples Republic of China) as the mainland was not recognized by the U.N. Neither could support their ally North Korea at the Security Council meeting.

Twenty-one countries of the U.N. made up 10% of the military while the United States made up 90%.

The Korean Conflict began on June 25, 1950, when North Korea invaded South Korea following clashes along the border and rebellion in South Korea.

North Korea was supported by China and Russia. South Korea was supported by the United Nations, principally the United States.

After John and Edith met, war broke out in Korea. Things weren't going well for the United Nations Forces during the first year of the war. John received word in September 1950 that he was being activated and would again be going to war.

Over 500 replacements and hospital returnees joined the 5th RCT in January 1951. Sgt. John Mullenax was among them. He was assigned to Company F to serve as a platoon sergeant. The new men received a 7-day orientation course. Then, on January 29, 1951, the 5th was ordered to attack northward to secure a critical mountain pass.

The second Battalion was assigned a long ridge dominated by Hill 256. A flight of P-51's kept the enemies' heads down as the companies commenced their attack toward the summit.

General Ridgeway, 8th Army Commander, was there the following morning when the 2nd Battalion reached the crest of Hill 256. He wrote: "In late afternoon, I personally visited the 2nd Battalion, LTC Ward, commanding, and went over the ground taken by E and F Companies. This operation achieved the true measure of tactical success – key terrain, a vital mountain pass – seized with heavy losses inflicted and only light losses sustained. The reason was due to proper appreciation and use of terrain and high leadership, whereby high-class infantry with supporting air and artillery worked its way along the ridges until all dominating ground was taken. This operation furnishes a fine example of how it ought to be done." Sgt. John Mullenax was back in combat.

The 5th RCT slowly ground north, reaching Seoul's outskirts on March 10, 1951, and the 38th Parallel twelve days later. It relieved the 21st Infantry on April 20, 1951, dug in near the town of Unjimal, and prepared to attack north to Kumwha on the following day.

Chinese artillery fire began falling on Fox Company at around 7:00 o'clock that evening. John received a fragmentation wound to his right hand, but not bad enough to impair its use*. Two hours later, the entire Chinese 20th Army attacked the United Nations line.

Shortly before midnight, Ridgeway's successor, General Van Fleet, ordered the 8th Army to pull back from the contested area and fall in on Line Lincoln, a few miles north of Seoul. The 5th RCT would serve as rear guard for the withdrawal. The third Battalion made it through the pass safely, but the 1st and 2nd Battalions and the 555th Field Artillery Battalion were ambushed, near Pisi-gol, by thousands of Chinese troops.

Sergeant Mullenax took charge and led his company and other surviving members of his platoon. The 5th RCT, on the other hand, lost 20 KIA, 289 WIA, and 243 MIA. A significant number of the latter would die in captivity from wounds, maltreatment, and starvation.

The following day, in a letter to Edith, John wrote: … Last night we fought all night and was surrounded, but got out. I am now back across the 38th Parallel Line… I led the Company out last night with my Platoon. Hon, I got to get some rest tonight. I didn't get any rest last night. I guess you are hearing about this over the radio… Well, Hon, there isn't much news…

On May 16, 1951, John wrote that he had been promoted to Master Sergeant and had been appointed first sergeant of Company F. He added that the 5th RCT was heading north again and that the Chinese troops, craving American cigarettes, were surrendering by the droves.

When John wrote Edy on June 25, 1951, Second Battalion was dug in on a secondary line of resistance, a backup resource to the frontline troops. He was feeling "tired but OK." He probably knew at the time that he would be leaving Korea in a month, but he didn't mention it. John said a final goodbye to Fox Company and the "land of milk and honey buckets" on July 26, 1951, and headed home.

Medals Awarded

1. Purple Heart with 1 Oak Leaf Cluster
2. Bronze Star
3. National Defense Service Medal
4. United Nations Service Medal
5. Combat Infantry Badge
6. Korean War Service Medal with 2 Bronze Stars
7. The Republic of Korea Unit Citation

CHAPTER FOUR

RESERVISTS ARE CALLED FOR TRAINING

LETTER 1 TO LETTER 88

Letter #1[1]
September 21, 1950

To: Edith Zinkhan
Frank, WV

From: Sgt. John W. Mullenax 33883656
"Co" H 2nd Bn 511 Air
Fort Campbell, KY

Dearest "Edy"

    Just a few lines to let you know that I arrived OK. I got here about 1 O'clock today. Darling, I want you to know how much I think of you. There were a lot of things I wanted to say but couldn't because it was so hard for me to leave.

    I will have to close for now. The sergeant said we had to go for clothes tonight. I will be here for 21 days, and where I go then, I don't know. They said most of my training would be after night. I hope this finds everything OK back there.

    Write soon.
    Love for always,
    Johnnie.

---

[1] *Arriving at Fort Campbell, Kentucky*

Fort Campbell 1950; Sgt. Mullenax is in back row, fourth from the left

Letter #2
September 22, 1950

To: Edith Zinkhan
Frank, WV

From: Sgt. John W. Mullenax 33883656
"Co" H 2nd Bn 511 Air
Fort Campbell, KY

Dearest "Edy"
    Just a few lines to let you know that I am getting along OK and hope this finds everything OK with you. How did the stove prove out? OK, I guess you had a time installing it.
    Did Betty and Howard[2] get up to see you? I sure would like to have been there.
    They sure have been pushing us through here. If you don't get mail regularly, don't feel bad about it, because we are going to train night and day. I have been thinking this will be the first weekend apart. I guess we will have to get used to it. I won't be able to get a pass from here. We are restricted to camps.
    Tell Bennie I said to be a good boy and listen to you and get his lessons every night. Darling, I have to close, for it is time for the lights to go out.

    Good night and sweet dreams
    With all my love,
    Johnnie.

---

[2] *Betty is Edy's sister*

Letter #3
September 26, 1950

To: Sgt. John W. Mullenax 33883656
"Co" H 2nd Bn 511 Air
Fort Campbell, KY

From: Mrs. H. W. Mullenax
Blue Grass, VA

Dearest Johnnie,

    I received your letter yesterday. Was so glad to hear from you and know you got there OK. I know it went just as hard for you to have to go as it did for us to see you go.

    Don't worry about us. I hope you will like it this time. Won't seem so long if you do. There were so many things I wanted to say to you before you left, but I just couldn't.

    They are getting along very well at the mill. They sawed yesterday. They will finish skidding at Bob's today. Virgil, Brooks, and Ben went to haul logs today. They didn't get much done last week for the rain.

    Ben paid the hands Saturday, guess he did it all right.

    Looks like the Korean War is about over. Hope you won't have to go there.

    Meade is still down at the hospital. Gladys was down Sunday. I think he is a little better, don't know how long he will have to stay.

    You didn't say what you were going to do with your car. Did they say anything about you taking it?

    We had three frosts, and its real cold here. Now the radio said it would warm up today; it will have to get at it if it's going to.

    Did you see the eclipse on the moon last night? I guess that was the cause of this cold spell.

    As news is scarce, I will close for this time. Write real soon. Be good.

Love,
Mother

I am now back across the 38th Parallel Line   Julie Mullenax Van Meter

Letter #4
September 25, 1950

To: Sgt. John W. Mullenax
"CO" H 2nd BN 511 AIR
Ft. Campbell, KY

From: Edith Zinkhan
Frank, WV

Dearest Johnnie,
    Just received your letters and you will never know how glad I was to hear from you.
    Hope you're feeling OK. We are fine, but lonesome for you. Betty, Howard and Pat came and the first thing Pat said was "Where is Johnny". They came Saturday evening and left this morning.
    They put the stove in Friday. I like it fine, but there is no water pressure in the kitchen.
    Honey, I have missed you terrible but, guess it's just the beginning. I'll sure will be glad when this is over. Howard said as soon as you were stationed permanently, we three will come and see you.
    I went to Elkins last Friday with Jack and Elsie and got a few groceries for the weekend. That's about the only place I have been since you left. The pictures came but, I haven't gotten them up to your Aunt Gerties yet. They are good.
    Yes, I got the record "Always" when I went to Elkins. If my letters don't sound right hon, it's because the other half of me is with you. Things just aren't the same without you. Just remember I love you very much and will be waiting for you.
    Ben has been very good, but mischief. There is no news hon, so I'll close for now. I'll write 2 or 3 times a week. Even if I don't get any from you. I understand hon, if you don't write regular.
    So, Sweetheart I'll sign off with all my love for you.

    Always Love,
    Edy

Letter #5

September 27, 1950

To: Sgt, John W. Mullenax
"CO" H 2nd Bn. 511 AIR
Fort Campbell, KY

From: Edith Zinkhan
Frank, WV

Dearest Johnnie,

    While Ben is getting his lessons, I'll write a few lines. How are things going? Not too bad, I hope? Things are awfully lonesome here. I have plenty to do but can't get started. Dewey came today and fixed the water in the kitchen.

    Elsie has been coming over at night for a while. Bennie's teacher "Hope" was here last night, and did Ben get to work! I sent those pictures up to your aunt by mail. For I hate to ask anyone to go anywhere for me.

    Ralph Stone and June Kesner will call, Ralph passed, but I never heard about the other.

    Hon, if this letter sounds crazy, I may have part of Bennie's geography in it, for he is asking me the questions right and left. I mailed your letter this morning. Hon, I don't see how we are going to stay apart, do you? But guess Uncle Sam has the answer to that. There isn't any news to write about. Oh yes, Howard and Ben gave three pups away Sunday, and Ben gave the other away today. That's over with, ain't you glad, Ha.

    Ben finally finished his lessons. Honey, I really miss you. Tomorrow will be Wednesday. We were supposed to have a date. Remember, guess that will be one of many. You won't be able to keep and it's been one week since I have seen you, but it has been much longer to me. Well, Hon, guess I'll say good night, sweet. I love you more than words can say.

    Goodnight, Sweetheart and have sweet dreams.
    Love you for always,
    "Edy."

Letter #6

September 28, 1950

To: Edith Zinkhan
Frank, WV

From: Sgt John W. Mullenax 33883656
408 QM Co. 11 AIR Div.
Fort Campbell KY

Dearest Darling,
    I have got a few minutes off. They are sure pushing us around here. I am transferred to Quarter Master Co. I am getting better chow here. I drove a truck all day. Darling, I hope you still love me. I love you and sure miss you. I haven't heard from you, but I know the reason, because of my moving around, the mail can't keep up with me.
    Honey, how did the stove prove out, fine? And I hope you like it. I just don't know how long I will stay here. I don't know whether I can get a pass yet. I will let you know. So, don't come down until I know for sure. I just ran out of ink and had to borrow from a buddy. Darling, I have a lot of work to do tonight yet. I get up at 4:30 o'clock; it sure keeps me busy. Darling, hold on; it won't be too long before I will get back, and if I sign up for the Reserves again, I want you to kick me good.

    "Write Darling"
    "Good Night, Honey"
    With all my love
    Johnnie

I am now back across the 38th Parallel Line   Julie Mullenax Van Meter

Letter #7

September 28, 1950

To: Sgt. John W. Mullenax 33883656
"CO" H2M Bn 511 AR
Fort Campbell, KY

From: Edith Zinkhan
Frank, WV

Dearest Johnnie,
    How are things for you by now? Better, I hope. Everything seems like a dream; I just can't get through my big head that we're apart. I miss you more every day. I hope you are shipped closer home.
    Ben and I are fine. He has gotten a few spankings since you left. I have received only 2 letters from you, hon. I know you said you wouldn't be able to write regular. I received both of them on the same day. It's terrible, hon when I can't see you or hear from you.
    I just heard tonight that Ted Mullenax's little boy and another boy were hit by a car. The car didn't even stop. They said the man was drunk, and his name was Oney Mullenax. Do you know him?
    I started this letter last night but ran out of news. Sweetheart, there isn't much more to write about until I hear from you. This will be another lost weekend. Betty was here last weekend, but you were missing. There is nothing, or no one could take your place.
    I'll sign off for now and get to work. Today is Friday, "cleaning day".
    I am close with all my love and hope you won't forget (little me) while Uncle Sam has you. Hon, in your other letter I forgot to put your number in it. Hope you received them. This is the 3rd one.

    All My Love,
    Always,
    "Edy"

Letter #8

September 30, 1950

To: Sgt. John W. Mullenax 33883656
408 QM Co 11 AR Div
Fort Campbell, KY

From: Edith Zinkhan
Frank, WV

Dearest Johnnie,
    Received your letter today and am sure glad to hear from you. I have been looking for one for days. Sweetheart, you know I still love you, and it's for good. Hope it will always be with you too.
    The stove is swell, and I'm glad I got gas instead of electric, for I think I like it better. I haven't been cooking much.
    Hon, if you didn't receive my first two letters, I forgot your serial number. Would that delay your letters?
    Hope, Ben's teacher was up again last night. Mr. Lambert cut all those low limbs off, and the trees look sorta bare. Boy Honey, you can't be out soon enough for me, for I sure do miss you. This is our second weekend apart, and each one is worse.
    I'll do worse than give you a kicking when you even look like you want an Army career if it means we are apart. I bet it's sure hard for you to roll out of bed at 4:30am.
    I bought me a coat, not a new one, but it sure is a nice one. It's a 100.00 coat for 25.00, and it s only been worn about 10 times. But you won't like the color, it's black, but it's a good coat.
    Hon, I won't even have enough money for a couple of months to come and see you. I won't be able to save much this month; the stove payments are 25.00 a month.
    Ben just finished his lessons. Well, Sweetheart, I'm running out of news, only I love you and need you, for I'm only half a person without you.

So, hon, I'll say good night, hoping you're OK and will be waiting with open arms and all my love in them.

>Good night Sweetheart, and Sweet Dreams
>All My Love for Always
>"Edy"

Letter #9

>October 1, 1950

>To: Sgt John W. Mullenax
>408 QM Co. 11th Air Div
>Fort Campbell, KY

>From: Edith Zinkhan
>Frank, WV

>Dearest Johnnie,

Here comes a few lines for tonight. I have just been down talking to Ruth. Then ran into Les and Hope, Sam and Ruth. They tried to get me to go with them, but I wasn't dressed to go anywhere, so I came home and thought I would write you a few lines.

Bill Hoover asked about you this afternoon. This has sure been a lonesome Sunday for me. I sure do miss you, hon; I will never get used to it. Elsie was over for a little while. Well, Sweetheart, I'm running out of news. So, I am going to say good night.

>With all my love for always,
>Your Love
>"Edy"

Letter #10

Oct 1, 1950

To: Sgt John W. Mullenax
408 QM Co 11th Air Div
Fort Campbell, KY

From: Benny Zinkhan [3]
Frank, WV

Dear Johnnie,
    How are you doing in the Army? I am doing fair in school. I hope you get out of the Army soon. I gave all my puppies away. Pat came up to see me.

    Love,
    Benny

Dad and Ben

---

[3] *Benny is Benjamin Christian Zinkhan III – He is "Edy" Edith Parsons Zinkhan's son.*

I am now back across the 38th Parallel Line   Julie Mullenax Van Meter

Letter #11

October 1, 1950

To: Sgt John W. Mullenax
408 QM Co 11th Air Div
Fort Campbell, KY

From: Mamie Mullenax
Blue Grass, VA

Dearest Johnnie,
    I received your letter Saturday, am so glad to hear from you and know you are alright. We are all very well.
    I was over at Gertie's last night a while. Mildred (Joe's wife), Joe (brother), Dad, and the children went to the movies, and I stayed at Gertie's. Gertie had a letter from Edith (sister). She sent her some pictures we had taken here Sunday. They were right, good, a little bit dim.
    Meade (Glady's sister, her husband). Came from the hospital Friday. He is some better, but not well, with his ulcers of the bowels.
    We got the check from Smith, $19.84 cents. They sold what cherry and walnut they had to the Wolf Company. They have it loaded on Virgil's truck, ready to go to Bartow in the morning. There are about four thousand of them.
    Tuesday, they take a load to Marlinton that good 4-quarter oak and maple they sawed all last week. But, the skidder had to stop to cut corn. They don't have any logs.
    I wrote to you last Tuesday. Have you got it yet? Hope you like your work. Joe (brother) thought he would come for the car Friday. Gene Rexrode (cousin) is talking of coming with him. Joe is going to see him.
    I will write and tell you if he is coming Friday or not. Your check from the Government came. Joe will bring it with him when he comes.
    Honey, I miss you, so I am hoping and praying you will get along alright and won't have to leave the states.

    Write real soon,
    Lots Love
    Mother

Letter #12

October 2, 1950

To: Miss Edith Zinkhan
Frank, WV

From: Sgt John W. Mullenax
408 AB QM Co. 11th ABM Div.
Ft. Campbell, KY

Dearest "Edy"

    Received your letter today. Wrote the September 17. I am sure glad to hear from you. I look forward to hearing from you. Honey, it is hard for me to get accustomed to this life. I don't think that I will be here long, in fact, I don't like this camp too well, but maybe because it is a strange place.

    I heard from Mother, and she said they were getting along OK with the mill. I sure will be glad to get back and get one of your meals off the new stove. Tell all the folks "Hello" for me.

    I got a permit to drive my car on the post. It cost me 5 cents with a lot of red tape.

    Tell Ben I said to get his lessons. Well, darling, I can't think of things to write about.

    Darling, I hope it won't be too long before I see you. Keep writing if you don't say but a few words.

    Goodnight Darling,
    Love,
    Johnnie.

Letter #13

October 3, 1950

To: Sgt. John Mullenax 33883656
408 QM Co. 11 Air Div.
Fort Campbell, KY

From: Edith Zinkhan,
Frank, WV

Dearest Johnnie,
    Here comes a few lines early in the morning. I went to Durbin last night. Elsie and I and the brats. I paid the store bill, and then we went to the PTA meeting at the schoolhouse.
    Bo Hiner has to leave on the 16th for Camp Usta, VA.
    I sure do miss you, hon; I'll never get used to it. It doesn't seem right to go anywhere without you. Guess I depended on you too much. But I'll sure be glad when this is over. Sweetheart, this is a short letter for I have to get Ben up. Hope I get a letter today from you
    I still love you, hon, more than I ever realized I could love any man. Hope you don't let me down in the end. Well, hon, I'll sign off with all my love. Write often.

    Love for always,
    "Edy."

    P.S.: Heard Joe had your mother and father up to your aunts over the weekend, but they never stopped.

Letter #14

October 3, 1950

To: Sgt John W. Mullenax, 33883656
408 QM 11th Air Div.
Fort Campbell, KY

From: Mrs H. W. Mullenax (this is John's mother, Mamie)
Blue Grass, VA

Dear John,
     I wrote you a letter yesterday. Didn't know when Joe was coming for the car. He has decided to come Friday. Harry Puffenbarger and Gene Rexrode are coming with him.
     Ben started to Marlinton with the lumber this morning.
     Hope you are alright.

     Love,
     Mother.

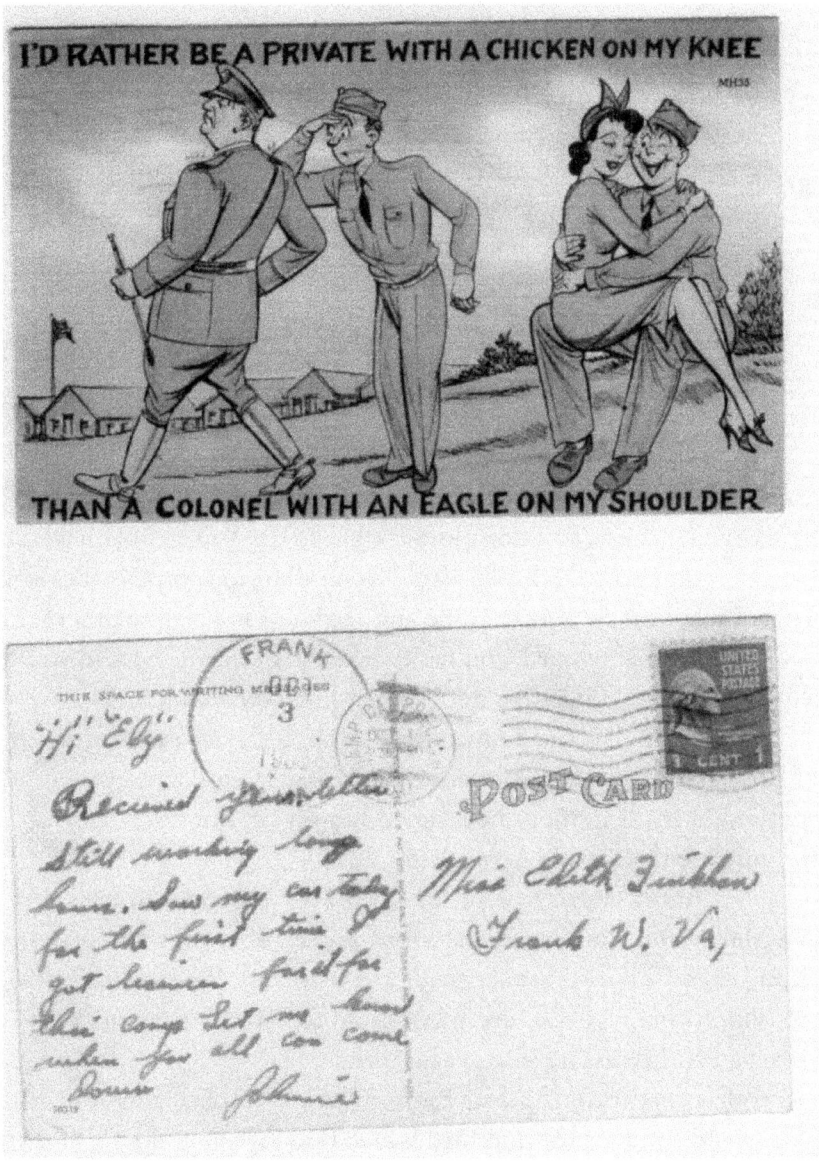

Funny postcard sent to Edith (She didn't understand it).

I am now back across the 38th Parallel LineJulie Mullenax Van Meter

Letter #15

October 4, 1950

To: Sgt. W. Mullenax 33883656
408 QM Co. 11 Air Div.
Fort Campbell, KY

From: Edith Zinkhan
Frank, WV

Dearest Johnnie,
    Received your card, and it puzzled me. It didn't sound like you. I have missed you so darn much; maybe it's me. Mom hasn't come and doesn't write like she is, so I don't know what to do. You know I have Ben and Johnnie, you know I'm broke also; I don't want to complain, but you also know I wouldn't be able to come and see you for a couple months.
    Johnnie, in your card, you never asked me to come or said you wanted me to come. Just let me know when I could come.
    You said you received my letter. You should have received several by now; I have sent 5 or 6. But, I received only 3 and 1 card from you. Guess you are tired of reading about how much I love you. But of course, it takes two to complete the case. HA
    I'm running out of news. Hope you will have more time next Sunday. Maybe you could write me a long letter. Hon, your letter means everything to me. That is just about all I have to look forward to; I have been to Elkins once and to Durbin last night to PTA. I just can't go anywhere; guess I'll just have to get used to it.
    Well, sweetheart, I'll close for now, hope you don't forget me completely and think of me once in a while. I do love you. So, I'll close with that, write often.

    All my love always,
    "Edy"

Letter #16

October 5, 1950

From: Edith Zinkhan
Frank, WV

To: Sgt. John W. Mullenax 3388365
408 QM Co. 11 Air Div.
Fort Campbell, KY

Dearest Johnnie,
    Received your letter and am sure glad to hear from you, even though you don't say you love me. Maybe you have forgotten about love. HA.
    As soon as you know when you will be for a while and after the first of next month, I'll come and see you if you want me to. I'm sure homesick to see you. Hon, I'm not worth 2 cents without you. I bet I have lost 5 lbs. since you have left. But nothing to worry about, you don't like me fat. HA.
    I picked almost a bushel of grapes. I'm going to make some jelly and can some juice, and I also wash today. Guess I will clean the house tomorrow; it sure could stand a good cleaning.
    Boy, Hon, I sure will be glad when you are back. I can cook you a good meal. Don't guess I haven't eaten a good meal or rather a whole one since you have left.
    Well, Sweetheart, I hardly know what to write. I could write and keep on writing about how much I really love you, but you know all of that.

    So, I'll sign off for now.
    With all my love for always
    Your future wife (I hope)
    Love,
    "Edy"

    P.S.: *Ben is sitting here figuring out how long it will be before you are home. It can't be too soon for me.*
    Good Night, Hon.

I am now back across the 38th Parallel Line   Julie Mullenax Van Meter

Letter #17

October 5, 1950

To: Edith Zinkhan
Frank, WV

From: Sgt. John W. Mullenax 33883656
408 QM Co. 11 Air Div.
Fort Campbell, KY

Dearest Edith,
 Received your letters and I sure was glad to hear from you. Honey, I have been hauling the reservists to their training area. I had to get up at 3:30 this morning. And to think I am an old sleepy head. HA.
 Darling, I got post tags for my car. I had a hard time getting them, but now that I have them, I would like you to come down to see me. I am not sure how long I will be here.
 Glad to hear that you got out. I don't want you to lock yourself up. Honey, I know that I can depend on you and when I get back, I want you to be the same as I left you. Hon, has your mother come up yet? If she does, I want you to come down.
 Darling, you don't know how much I miss you. By the way, did you get the mixer yet? Tell the Lamberts hello. I don't have time to write much, when I do, I will write to you. Because you are the one, I want to keep in contact. Honey, don't worry about me because I don't have time to ever get out.
 I love you, and don't forget it.

Love for always,
Johnnie.

P.S.: If you can't get the money, I will get it for you to come down

Letter #18

October 6, 1950

To: Sgt John W. Mullenax
408 A B QM Co. 11th ABN Div
Fort Campbell, KY

From: Edith Zinkhan
Frank, WV

Dearest Johnnie,
    No letter today, so I'll scribble a few lines. How are things going? Fine, I hope. I think time just stands still, since you have left. Each day seem to be getting worse for me.
    Hon, on the card you wrote that you have licenses for that camp. Does that mean you will be there for a while? You never said. I received a letter from Betty today, Howard had been laid off for a week. So, I won't be able to depend on them to bring me.
    If and when I come to see you, that is when you want to see me and when I get the money, I'll leave Bennie over at Lambert's.
    Hon, I received a $37.00 increase in my checks. I have cut down on lots of things such as 5 & 10 stores. Ha. Can you guess why? Because Johnnie's not here to take me.
    I would give anything if you were just near me. Tomorrow is the opening of the Forest Festival. Remember, last year and the car broke down. Have you been getting my mail? Maybe that is why you haven't been writing.
    There is not much that I can write about when I don't hear from you. I love you Hon, hope you don't forget how much I do love you. But of course, if you should find someone you love more, just let me know. Well, Sweetheart I'll sign off for now. Please write me a long letter and let me know some of your feelings, I have poured out mine. Guess it's because I have been so darn lonesome for you. If you should get tired of reading about them tell me.
    I'll sign off for now. Excuse the misspelled words.

    All my love,
    For always
    "Edy"

Letter #19

October 7, 1950

To: Sgt John W. Mullenax 33883656
408 A&B QM Co. 11th ABN Div.
Fort Campbell, KY

From: Edith Zinkhan,
Frank, WV

Dearest Johnnie,
    Here come a few lines, didn't get a letter today. I cleaned the house today, the first since you left. Bet you're saying, "Gosh, she's getting lazy", and that's the truth. I have no interest in anything since you have left.
    I think everyone went to the Forest Festival today. Received a letter from Mom today. She said she would come up if I sent her the money, and I don't have it until the first. I am not cashing the checks that I have saved to come and see you.
    Lamberts said they would keep Bennie for me any time; I wanted to come and see you. After the first of next month, I'll have enough money, I think. I have to get the darn coal yet. Elsie got her driver's license today. The stove had a leak in it, but Dewey came and fixed it.
    Well, Sweetheart, there is no news.
    So I'll sign off.
    With all my love, Good night and sweet dreams (not wet ones) HA.
    for Always
    "Edy"

P.S.: Floyd's son Paul received his call.

P.P.S.: I'll make you a box of candy if you will be there long enough to get it.

Letter #20

October 7, 1950

To: Edith Zinkhan,
Frank, WV

From: Sgt. John W. Mullenax 33883656
408 QM Co. 11 Air Div.
Fort Campbell, KY

Dearest Darling,
    I have been receiving your letters regular now. Darling, I sure love to hear from you. It goes harder for me to be in this Army, this time more so than the other time, on account of you. Have you got your coal yet, and do you have enough wood? About the card, don't take things to heart. I say, in fact, I am so rattled I don't know half what I write.
    You know I want you to come down. I don't know what is wrong. We haven't been apart long, but it seems as if it was a year. Honey, you know I love you, and I am a straight shooter; as long as we have been going together, I never went wrong. I have to go tonight and exchange some clothes they gave me. They're too small.
    I don't know whether you can read this or not because I am writing this in my bunk, and the boys are always jumping around.
    Honey, I don't know how long I will be here but not for long. I don't think. Well, honey, I will have to stop for this time. By the way, the boys look at your picture and go wild.
    Good night Darling, and our love keep us together no matter how far apart we are.

    Love for Always,
    Johnnie

    P.S.: I love you.

Letter #21

October 9, 1950

To: Edith Zinkhan
Frank, WV

From: John W. Mullenax 33883656
408 QM Co Air Div
Fort Campbell, KY

Dearest Darling,

    I received two letters from you today, and I am sure glad to hear from you and to know that everything is alright. I have today off, and I'm sure going to catch up on my sleep. They sure are running us here day and night. You asked if I was going to stay here awhile. Honey, I wish I knew; I am sure to ship out any time.

    Darling, you asked when I wanted to see you anytime, and it can't be too soon because Hon. I miss you. Darling, have you got the coal yet? If you can't get it, I will let Brooks know, and he can take the company truck and get it. I don't want you to freeze this winter.

    Darling, the more you tell me you love me, the better it makes me feel because I love you. You said if I found another, well, don't worry, I am not looking for another. There is only one for me, and you are the only one who knows who it is?

    How is Ben getting along in school? Tell him I said to be a good boy. Darling, I sure would like to be back there today because this is going to be a long Sunday without you. Well, I will be looking for a letter tomorrow from you.

    Darling, don't worry about me because I still love you and always will.

    Love,
    Johnnie.

Letter #22

October 9, 1950

To: Sgt John W. Mullenax
408 QM Co 11th HB
Fort Campbell

From: Mamie Mullenax
Blue Grass, VA

Dearest Johnnie,

    Received your letter. Was so glad to hear from you, and I know you are getting along alright. We are very well. They didn't saw any last week, didn't have enough logs skidded. They hauled six loads. Don't look like they are going to get much done this week, for it's raining this morning.

    Maybe they will give you a furlough to bring your car home; I hope they will. We heard over the radio that Truman was only going to keep the reservist for a year. I hope that is so. Bob Simmons (He is a fellow Highland County VA native) is missing; they hadn't heard from him for a month. A man came out from Staunton and said he is missing.

    We don't have any radio now, the tubes went bad, and we haven't been able to get any. Tried to get them in Staunton. Dad (John's dad) doesn't know what to do without it. We don't know how the war is going.

    We went over to Staunton last week. Dad wanted to get tires for the Company truck, and I went to see Edith (sister). She is just about the same. Fred (brother) was here Sunday. Guess I won't be going over to Gertie's so often since you are gone. I am mailing your checks. It said not to forward to another address, so I thought I would put them in another envelope and send them. I guess you hear from Edith (this is Edy, his girlfriend) often. I am trusting everything will turn out for the best.

    Lots Love
    Mother

I am now back across the 38th Parallel Line   Julie Mullenax Van Meter

Letter #23

October 9, 1950

To: Sgt John W. Mullenax 33883656
408 QM Co 11th Air Div
Fort Campbell, KY

From Edith Zinkhan
Frank, WV

Dearest Johnnie,
    Received 2 letters from you today and sure glad to hear from you. Ben sure was tickled to get your letter.
    Hon, I would love to come and see you. I haven't heard from Mom yet, since she wrote for the money to come on. You be sure you will be there when I do come. Hon, don't you worry about me not being the same as you left me, for I sure will be. You are the one and only for me, and I sure will be glad when you are back to stay.
    I finished up last week's ironing today and made some more grape juice.
    Hon, I won't have the money to come down until after the first. Think, Hon, you have been gone 3 weeks; it has seemed forever to me. I can't seem to get it straightened out. You better be a straight shooter, ha. You know I will be hon, for I love you. It looks like I have taken to be a sleepy head now, instead of you, for I sleep all my spare time, but I'm not getting fat. Ha. If anything, I'm thinner, but I imagine that because I haven't been drinking any "Old Hickory," don't you. No, I haven't got the coal yet. Think I will try and get someone else to get it. The wood is holding out. OK. It hasn't been very cold.
    Your Sis Esta ran all of the Davis's off Saturday. Guess she finally got fed up with them
    Well, Sweetheart guess I have told you all the news I know about. So, I'll sign off for now. I love you, Hon, so that is all for now. Your letters were sweet; you better mean what you say; ha, I am just kidding.

    All My Love for always
    "Edy"

Letter #24

October 10, 1950

To Edith Zinkhan,
Frank, WV

From: Sgt. John W. Mullenax 33883656
408 QM Co. 11th Air Div.
Fort Campbell, KY

Dearest Darling "Edy"
    Will try and write a few lines tonight to let you know that I still think of you and can't get you off my mind. Darling, I am going to try and get a 3-day pass soon, that is, if I am here long enough to get one. If I do, I will be home. I haven't had the wet dream yet, but I look for it anytime, Ha.
    Darling, I got a letter today mailed the 7th.
    There were 4 of us who went down to Nashville Sunday to the stock car races. Boy, was it good to get out of this place for a few hours. My face is sure catching, hell, I have to shave every day. By the way, they told me I had to shave off my mustache. You should have seen me; I look like hell. So, I'm growing it back now. The Captain gave me a hard look the other day. I'm going to wear it until I get direct orders to shave it off.
    Honey, the rumor is we may stay here until December. The rest of the boys that came in, ship out to Fort Dix and then to Germany I don't know what they are going to do with me. I haven't been doing anything but driving a truck here. I got to get my laundry ready to go out in the morning.
    Well, Darling, let me know how you are getting along and the news.

With Love Forever,
Johnnie.

Letter #25

October 11, 1950

To: Sgt John W. Mullenax
408 QM Co. 11th Air div
Fort Campbell, KY

From: Mamie Mullenax
Blue Grass, VA

Dearest Johnnie,
 Just a line to let you know I received your letters. I am always so glad to get them. It's rained for 2 days, but it looks like better weather this morning. Dad got a radio. The one they had down at Harry's, it's a good one. (Harry's is a store named Puffenbarger's Mercantile)
 Virgil said he hadn't gotten paid for 2 loads. He hauled one load of lumber to Petersburg and one to Wolf's in Bartow. Do you know anything about it?
 Ben (brother) is looking after the books, alright. I think he spends a good bit of time on them. He made out the men's insurance.
 Joe's turkeys went away yesterday. Don't know if he made anything on them yet or not. I can't think of anything to write. Be good and answer soon.

 Love,
 Mother

Letter #26

October 10, 1950

To: Sgt John W. Mullenax
408 QM Co 11th A B Div
Fort Campbell, KY

From Edith Zinkhan
Franklin, WV

Dearest Johnnie,
    Here come a few lines hoping they will find you OK. I received a letter Monday. But haven't received any since. Hon, I'll never get used to being without you; I miss you more each day.
    There isn't much news to write about. Ben and I have been having a few rounds about his schoolwork. His teacher says he can do much better than what he has. I have my washing and ironing done up for once. I haven't gotten my coal yet.
    Hon, when I come down, you're going to have to tell me how to get in contact with you or meet me.
    Guess Mom will be here sometime next week.
    Well, Hon, maybe I can write some more now, I just took a bath, and I feel better, but still lonesome for you. We have had some real nice weather this week.
    Well, Sweetheart, I hope I get a letter tomorrow; I am running out of news; I love you, Hon, but you know that, don't you. So, I'll sign off for now with all my love.

    For always,
    Love "Edy"

Excuse the pencil and paper.

I am now back across the 38th Parallel Line   Julie Mullenax Van Meter

Letter #27

October 10, 1950

To: Sgt John W. Mullenax 33883656
408 QM Co 11th Air Div
Fort Campbell, KY

From: Edith Zinkhan
Frank, WV

My Dearest Johnnie,

    Here comes a few lines hoping they find you OK. It sure has been a dreary old day here, rained all day long. Hon, I received a letter from mom. Says she will come as soon as I send her money to come on and enough for a dress. If I send her what I have, I won't even have enough left from next month to come and see you. She makes me so darn mad. Guess I shouldn't be telling you my troubles. You have enough of yours.
    I sure do miss you, Hon, on these rainy days. Well, I miss you, period. If I send mom the money, I'm coming down, or where ever you go and get a job. Is it OK? I can't sit around here going nothing. At least we will be near each other.
    Well, Hon, there isn't much to write about, only I love you with all my heart. So, I'll sign off, with all my love for always,

Love, "Edy"

    P.S.: *The Lamberts said Hello.*
    P.S.S.: *I haven't heard a thing from the mixer.*
    P.S.S.S.: *Floyd's son Paul got his call, they're sending him to San Fransico, Ca.*

Letter #28

October 11, 1950

To: Sgt. John W. Mullenax
Fort Campbell

From: Ben Zinkhan
Frank, WV

Dear John,

    I hope you get out of the Army soon. And you start writing to me when you have time. How is it up on the bunk?

    I like to go to school. The Yankees are in first place, and the Yankees are my favorite baseball team.

    Love
    Ben Zinkhan

Letter #29

October 12, 1950

To: Edith Zinkhan
Frank, WV

From: Sgt John W. Mullenax
408 QM Co. 11th AB Air Div.
Fort Campbell, KY

Dearest Darling,
    Received a letter today and I am glad to hear from you. I sure would like to be there to fill that seat at the movies. I haven't been to a movie since I have been down here. I wrote for them to come for the car, for I thought I was shipping out, but I don't know when it will be now. But, when I do, I get 10 days delay in route. I want to get back and eat some of that grape jelly.
    Honey, I don't want you to worry about me because I have told you that you were the only one and how much I love you. I don't see why you haven't been getting my mail. I have been writing every day or two.
    I have been putting a lot of hours in, but I am not working too hard. Darling, I hope you start getting my letters. I missed one from you yesterday. The boys went crazy over your picture. I told them I didn't want to catch them swiping it.
    Honey, when I do get home, you and I will spend every minute together. This is the longest we have been apart for 3 years. It seems as though it has been a year.
    Well, Darling, I can't think of much to write, so I will sign off for now. Darling, I love you, and I can't see why you doubt me.

    Love for Always,
    Johnnie.

    P.S.: *Send some stamps.*

Letter #30

October 12, 1950

To: Sgt John W. Mullenax
408 QM Co, 11th Air Div
Fort Campbell, KY

From: Edith Zinkhan
Frank, WV

Dearest Johnnie,
    Received your letter and it always makes me feel good to hear from you.
    Hon, I'm getting the coal soon from Bud Lockrage, 7 ½ tons. Kinsiner kept putting me off.
    Hon, I would be down there, but I don't have enough money to come. Hope you aren't moved farther away. I would love to see you. It's been 3 weeks today, and after the first week of next month, I'll see you. I hope. It seems like forever since you left.
    I washed today and fixed my coat. I'm really getting lazy. Mona and Bill don't come around anymore; she was up one time since I got my stove, and she didn't have much to say about it.
    Everything here is the same as ever "dead," or maybe it's me. They say Ralph Stone has really been on a drunk since he received his orders.
    I received another letter from Mom. I didn't send her the money, and I'm not going to. For when I come to see you, I'll leave Ben over at the Lambert's. He's not getting along too well in school.
    Well, Sweetheart, I'll say I still love you. Guess you know that. I miss you terrible, write often.

    All My Love for Always
    "Edy"

Letter #30 (continued)
    Ben has gone to church; they're having revival here and thought I would scribble a few more lines. Hon, do you think you will be lucky enough to stay in the states. Boy, I hope so.
    Well, Hon, Ruth sent for me to come down, and I just got back. She wants me to go to Elkins with her and Floyd tomorrow. So, I will have to put my wool up. So, sweetheart, I'll say good night and sweet dreams; I love you very much.

    All my love
    "Edy"

I am now back across the 38th Parallel Line　　　　　Julie Mullenax Van Meter

Letter #31

October 13, 1950

To: Sgt. John W. Mullenax
408 QM Co 11th Air Div
Fort Campbell, KY

From: Edith Zinkhan
Frank, WV

Dearest Johnnie,
　　　　Received your letter and it tickles pink to hear from you. Hon, don't worry about me; you know I'm not in need of anything but you. I will always need you, and I sure will be glad when the time comes that we will be together for always.
　　　　Boy, I sure hope you get a pass. If you do, stay there; I'll be down. For I sure would love to see you without your mustache.
　　　　Hon, I'm glad you got to go to Nashville, just so there weren't any blondes. Ha. Or redheads.
　　　　I hope you won't have to go overseas; tell the Captain they can send you home if they want. Ha. Well, Hon, I went to Elkins today with Floyd and Ruth.
　　　　Guess how much I weigh (126 lbs.). I weighed 131 when you left; that sounds good, don't you think.
　　　　Elsie was over for a while tonight. Ben is taking his six weeks test. Arithmetic 80, Spelling 88, and History 100. Not too bad, but it could be better. He hasn't finished the other subjects.
　　　　Hon, I still miss you and always thinking of you, and I love you lots.
　　　　Do you remember ever saying that you wish I would love you as you love me? Well, I do, Hon. Are you sorry? Pal has been barking her head off, and Ben has gotten him another cat. He worked all day last Saturday for Billy Tracy and made $50.00, and helped Sammy one evening and got another $6.00, and another evening for Bill Tracy, got $5.00.
　　　　He says when you come home, you and him are going to buy a farm. Two beats one, don't you think? Well, hon, I'll sign off for tonight, for my news are scarce. I love you, sweetheart, and good night and sweet dreams.
　　　　Love for Always
　　　　"Edy"

Letter #32

October 14, 1950

To: Sgt John W. Mullenax
408 QM Co 11th Air Div
Fort Campbell, KY

From: Edith Zinkhan
Frank, WV

Dearest Johnnie,
    Received your letter today. This past week I have received your letters regular, but I didn't before.
    Hon, I don't doubt you; I mean to sound like I do, guess I just get the blues.
    Boy, I sure will be glad when you get home; you will never know how I miss you. I have one heck of a cold, and I can't talk, guess that's a good thing, Ha. Guess it would be OK.
    If I looked like the picture, don't you think.
    Well, Soldier, I am running out of news; I went down and helped Ruth today. She is expecting her son and family. I'm also making her 2 apple pies; they sure look good. It has turned cold the last couple of days. Hon, I get my coal Monday or Tuesday. He had a load yesterday, but it was my run. I want a lump; maybe it won't smoke so bad. I have to have all new furnace pipes put in. I have it ordered.
    Well, Sweetheart, Enclosed are stamps. I'll sign off for now.

All my love always, "Edy."

P.S.: *I love you.*

Letter #33

October 14, 1950

To: Edith Zinkhan
Frank, WV

From: Sgt John W. Mullenax
408 QM Co, 11th Air Div
Fort Campbell, KY

Dearest Darling,
    Will write a few lines to let you know that I am getting along OK. I'm still thinking of you. I have to get ready for inspection tomorrow, I am sure shining shoes. Honey, I don't know how long I will be here, but I still would like to see you. But I wouldn't try to get a job here. We can't find out how long we will be here. One of the boys that came down is still with me (Hise).
    Honey, I hope I don't have to pull the whole twenty-one months. Twenty-one months is too many for me. Darling, I have your picture on the shelf where I can lay in bed and look up at you. I am sure in love with that person.
    Honey, I pause and think over the good times we have had together and hope we will have some more soon. I don't know what to tell you about your mother, but if she did come, you would be free to come to see me anytime. Honey, I don't want you to neglect things at home on the count of me. Hon, I hope these few lines find you well and everything OK. Tell Ben Hello for me.

    Good Night, Darling
    Love for always
    Johnnie.

Letter #34

October 16, 1950

To: Sgt John W Mullenax
408 QM Co. 11th A B div.
Fort Campbell, KY

Dearest Johnnie,
    Here goes a few lines. How are things going? Things are the same around here. Tonight, Saturday nights are the worst of all. Guess I will get used to it sometime, within the next 21 months, I hope. Guess I'll stay home tonight. I don't have enough energy to get ready to go anywhere, rather to the movies.
    Ruth and Floyd are going to Huntington sometime next week, and they're going to bring mom back with them. Ben and I went over to Lamberts for a while and ate hot dogs. They were really good.
    Well, Hon, there is no news only. I love you lots and miss you terrible.
    So, I'll sign off for tonight, I didn't receive a letter today, and I sure did miss it. So, goodnight, Sweetheart. I'll write a few more lines tomorrow.

Sunday
    Today is another lonesome Sunday; I think I have read everything in the paper. Well, Sweetheart, I hardly know what to write about, wish you were here; it sure has been a beautiful day. Everyone is out, and the trees are really pretty.
    Hon, I am going to sign off for now

With all my love
For always, Love "Edy."

P.S.: Hon, Ralph Stone was just here and got address and said to tell you he would see you next Monday or Tuesday.

Love
"Edy"

Letter #35

October 17, 1950

To: Sgt John w Mullenax
408 QM Co, 11th A B Air Div
Fort Campbell, KY

From: Edith Zinkhan
Frank, WV

Dearest Johnnie

    Received your letter today mailed Saturday. Everything here is OK. I have been working for a change today. Iron and 8 pts of grape jelly and canned 2 quarts of juice.

    Hon, if you're still there by the first, I'll be down to see you. I would be there now if I had my way. Mom will be here in 2 weeks.

    I didn't know just how to take the part of your letter, that said, not to neglect things on account of you. Guess I have been neglecting things since you left, but it's only because I have nothing to look forward to, but not because of you. Things aren't the same without you.

    Hon, I often think of the good times we have had and the times I was so mean and was sorry afterward, but there were times when I thought you didn't want me. I mean for always the right way, but I'm hoping I was wrong.

    Well, Hon, I ran out of paper and went down to Floyd's and bought some. Excuse the lines. I'm having a time with Ben and his lessons.

    I sure hope you don't have to stay the full 21 months, and we can get together for good. Hon, I won't be so hard to please, just so we can be together, that will be the most important thing. We will manage, don't you think? I stayed home most of the day, and the neighbor got worried and thought I was sick. I think I better stay home a little more, don't you think, Ha. When I go down to Ruth's, I never can get away. Well, Sweetheart guess I have written just about all the news. I'm thinking of you all the time and loving you too. So, I'll sign off.

I am now back across the 38th Parallel Line   Julie Mullenax Van Meter

    With all my love
Good night Sweetheart, and Sweet dreams.
Always, Love "Edy"

    P.S.: Did you receive the stamps? Anything else you need, just let me know.

Letter #36

October 17, 1950

To: Edith Zinkhan
Frank, WV

From: Sgt John W. Mullenax
408 Q M Co 11th A B Div
Fort Campbell, KY

Dearest "Edy"

How are you by this time? Fine, I hope. I'm getting along OK, but still, I don't like it here. Monna sent me the paper, and I am enclosing the clippings.

Darling, I am trying hard to get a 3 days pass so I can come and see you. I know how much we mean to each other since we have been apart for a while. I received the stamps today; I sure needed them "thanks a lot." I got one of your letters that was mailed on the 28th. It must have taken the long way around to get here.

Honey, I never will forget meeting that beautiful little woman at Propst Place at Bartow. You are the only one who knows what I mean. I sent my shirts to the laundry here, and they look like Hell.

Honey, I will be glad to get back and get one of the good meals cooked by the best cook that I know of.

Honey, I have missed writing you a couple of days. Don't let that discuss you. Write me every day. I go to mail calls every day looking for a few lines from the sweetest darling I know.

Well, Darling, I will have to close for this time. I have to get the platoon up at 4:30 in the morning. I can see me getting you up at 4:30. I bet I get a shoe over the head.

I love you, darling good night and sweet dreams.
Love,
Johnnie

Letter #37

October 17, 1950

To: John W Mullenax
408 QM Co 11th Air Div.
Fort Campbell, KY

From: Mamie Mullenax
Blue Grass, VA.

Dearest Johnnie,
    I received your letter yesterday. You don't know how glad I was to get it. I didn't hear from you last week. Thought they had shipped you off to Korea. I am so glad you are still there.
    We are very well. Dad is complaining of a cold. He came home last evening about all in, but he went again this morning.
    He brought the tractor over and fixed the road on Strait Creek. They are hauling logs from over there. Don't have any logs skidded back of the mountain. They hauled five loads yesterday. They are sawing now. They didn't get much done last week. I wrote to you about Virgil saying he didn't get paid for 2 loads, one of ties to Petersburg and a small load of Chestnut to Bartow. Do you know anything about it?
    They got word Bob Simmons was killed on September 10, 1950. Esther Hammer is still in the hospital. She is worse off. She had a stroke.
    Milton Mullenax (nephew, Joe's son) said to tell you to come home. If they don't know what to do with you, maybe they would give you a furlough in a month or so. But don't come without permission. I hope the war in Korea will soon be over, and you won't have to leave the states. I can't think of anything more to write.

    Be good, and lots of Love
    Answer soon.
    Mother

Letter #38

October 18, 1950

To: Edith Zinkhan
Frank, WV

From: Sgt John W. Mullenax
408 QM Co, 11th Air Div
Fort Campbell, KY

Dearest "Edy"

    Here comes a few lines, Darling, to let you know how much I love you, of which you know and how much I miss you. I am on C. Q. tonight, I started at 6 this evening, and I get off at 7 in the morning. Honey, I have your radio down here tonight with me. It sure comes in good tonight because I'm here by myself.

    Darling, I had a dream about you last night. You can guess what happened. Ha. I'm working on that pass again today. They talk like I may get it. I hope so, but I wouldn't be home but one night. One night with you means a lot to me, so keep your fingers crossed.

    I hope I don't come home at the wrong time, you know what I mean.

    Honey, this army isn't for me. I don't like it here. It is run by regular army men, and a reservist hasn't a chance here. If I get shipped to another camp, I may like it better.

    When I get some money, I am going to send some to you, so you can be near to me. Money means nothing to me without you. I made out a Class-E allotment of $50.00 to put in the bank.

    How is Bennie making out with his schoolwork? Tell him I have a watch for him if he passes his school work, and I want him to listen to you. I haven't found what I want to get you, Darling. I hope you get your coal; if you didn't, I would see that you get it.

    Honey, I don't want you to get caught without coal if they strike. I have heard from my mother, but she didn't write much about the business, so I guess it is coming along OK. Darling, there is much to write about here, only I am getting in long hours; if the government paid me by the hour, they would go broke.

Honey, I will stop for this time, hoping to get a letter from you tomorrow. I didn't hear from you today. Honey, I hope I have shown you how much you mean to me in the last few years.

Love,
Johnnie

P.S.: *You have my heart and mind with you always.*

I am now back across the 38th Parallel Line   Julie Mullenax Van Meter

Letter #39

October 19, 1950

To: John W Mullenax 33883656
408 QM 11th Air Div
Fort Campbell, KY

From: Edith Zinkhan
Frank, WV

Dearest Johnnie,
    Here's goes a few lines for tonight. Ben's in bed, and the radio is making a lot of noise.
    We went to the movies and saw a bang bang. Hon, I was lost without you, your seat was there, but it was empty.
    I made about 20 jars of grape jelly and have more to make. Hon, I have received one letter each week. Hon, I guess I am selfish, wanting you to feel the same as I was before I got so darn lonesome.
    I'll finish this tomorrow, Hon and go to bed, for I can't think straight tonight. I have been working the last week, but it doesn't help. Guess I am just in an awful mood tonight. So, goodnight, Sweetheart, I love you.
    Sunday – Here comes a few more lines. It's another lonesome and blue Sunday. Ruth Collins and I drove up to see your Aunt Gertie for a while. She told me you had written for someone to come after your car and then wrote not to come. Can't you keep the car? She said Mona and her family went to the Forest Festival, also Joe and, I guess, Mildred.
    Hon, I still can't think of much to write about. I have written in all my letters how much I love you. But you hardly ever mention love in yours. I used to think you really cared for me, but the last couple of months you were home, you had me puzzled. Maybe it's me. I know actions are supposed to speak for themselves.
    Hon, I have written every day this week, but if you don't write a little more often, it's the only thing I have to look forward to. Well, Hon,

I'll sign off, for I know this is a terrible letter. Please write as often as you can.

>All my love,
>"Edy"

>**P.S.:** *I saw Ralph Stone's name in the paper. He leaves soon. Excuse paper and writing.*

Letter #40

October 19, 1950

To: Edith Zinkhan
Frank, WV

From: Sgt John W. Mullenax
408 QM Co. 11th AB Div
Fort Campbell, KY

Dearest Darling,
    Will write a few lines, dog tired. I got up at 4:30, and it is 12 o'clock. I read your letter out in the woods. A Jeep driver got it for me. Honey, I know what you are going through due to the fact I miss you so much. Live on hopes, Darling. I hope to see you soon. If I can get a pass, I am going to bring you back with me. Do you think your Mother can take care of things there and Bennie?
    Well, Darling, it makes me feel a lot better in this army to know that you love me. So I will do what is right and get out as soon as I can. Some of the things they dish out here are sure hard to take from some of these jokers when they haven't had combat time or as much time in the army.
    Well, Darling, I will close for this time, hoping that my future wife will make out OK.
    Lots of love, and may we share that love together.

Love,
Johnnie

Letter #41

October 20, 1950

To: Edith Zinkhan
Frank, WV

From: Sgt John W. Mullenax
408 QM Co, 11th A B div.
Fort Campbell, WV

Dear "Edy"
    How is my darling getting along by now? Hope these few lines find you OK. Honey, you said you had nothing to look forward to.
    Honey, I haven't tried to discourage you. Because I think both of us have a lot to look forward to. Honey, you said about being mean to me. It wasn't your fault because I did get out of the way sometimes. You had a right to straighten me out. Honey, you are more than just a girlfriend or sweetheart to me. If you love me, you will know what I am saying.
    Honey, is Ralph Stone coming here? I was watching them parachute out of the planes today. One boy's parachute didn't open; he didn't know what happened.
    Do you think I should join them? They pay $50.00 a month more. This doesn't happen often; I mean, the parachute opens. No, Honey, I don't want to join them; all I want is to get out and back to you because I want to look after you and Bennie.
    Darling, I will close tonight, hoping you have faith enough in me to trust me because it is going to stay the same as it has the last 3 years.

    Goodnight Darling,
    Love,
    Johnnie

Letter #42

October 20, 1950

To: Sgt John W. Mullenax
408 QM Co. 11th AB Div
Fort Campbell, KY

From: Edith Zinkhan
Frank, WV

Dearest Johnnie,
    Here comes a few lines, hoping everything is in tip-top shape. I finally received a letter from you.
    "Say," Bud, what's wrong? You're slipping, you wrote last Friday night, and there were no more letters until Tuesday. Rather Monday night, you know how dead this place is, and when I don't hear from you, everything is dead, period, even me. Maybe I'm expecting too much. Guess I always do.
    I cleaned house today and made a cake, but it was a flop. You were talking about my cooking. I haven't cooked a good meal since you left. I read the clippings you sent.
    Hon, why do you never talk about our future? You have never made me feel sure of things for us. Is it because you weren't sure of yourself? Or why? I have all your shirts done.
    Well, Hon, I sure hope you get a pass. After mom comes, I'll have a chance to come down if you want me too. How much money do you think I will need? I don't think I will have, but about 50.00. I would love to see you and see how you look without that mustache.
    Well, Sweetheart, I'll sign off and say good night, and I love you very much.

    Good night and sweet dreams
    Love for always
    Edith Your battleax

    P.S.: *Now, this is news, I just received a summons to appear at Hope Mallow (Ben's teacher) and Wimer's trial. She whipped one of his boys, Charles. Maybe that happened before you left, did it? And I asked the boy on Monday to see his blue spot, and he said he didn't have any. So that gets me messed up in the darn thing, and Ben is also summoned.*

Letter #43 [4]

October 20, 1950

To: Sgt John W. Mullenax ER33883656

From: 408 QM Motor Pool
11th Airborne Div.
Fort Campbell, KY

The following named EM requests a seventy-two-hour pass from 0600, October 21 1950, to 0600, October 24 1950.

REASON: Man claims I owe him money for trucking that was done in the past. My brother has taken over my books, and cannot find the receipt.

        Sgt. John W. Mullenax ER33883656
        Cpl. Charles E. Dodson, 4th Truck Platoon Sgt.
        Sgt. Blanton

Disapproved
    Sgt. Mullenax not considered for 3 days pass at this time – for the following reasons.
    1. Received Summary Court Martial, October 14, for being drunk.
    2. Company Policy for a 3-day pass requires all requests to be in the orderly room on or before Thursday of the weekend pass is requested.

    T. H. Hudson, Capt.

---

[4] *Motor Pool Pass/ Denied for being drunk*

Letter #44

October 21, 1950

To: Sgt John W. Mullenax
408 QM 11th AB Div.
Fort Campbell, KY

From: Edith Zinkhan
Frank, WV

Dearest Johnnie,
    Received your sweet letter and am sure glad to hear from you. It's your letters that keep me going. Hon, I never knew what real love was until I met you, and there are no words that can replace what you mean to me, and no one could ever take your place with me.
    I sure hope you get your pass, and Hon, if mom is here, I'll go back with you. I'll have my fingers and toes crossed. Hon, the 28th is the bad luck date, you know what I mean.
    Hon, don't worry about getting me anything. You, yourself, will be enough for me. Ben was tickled when I told him about the watch. He is not doing too good in school since that trouble started. I think Hope is letting them get by with anything. I sure dread going down tomorrow. Just think, I never was mixed up in anything and have to get into this.
    Everything here is about the same. Everyone is talking about the trial. The Wimer's don't have a chance. He sued Hope for $300.00, I think. Just about everyone around here is summoned.
    Ben and I were just around trying to catch a mouse, but it got away.
    Well, Sweetheart, I love you with all my heart, and I'll say good night and give Ben a bath, and I have to take one. I'll have to sign off with

    Love for Always,
    "Edy"

Letter #45

October 21, 1950

To Sgt John W Mullenax
408 QM Co, 11th AB Div
Fort Campbell, KY.

From Mamie Mullenax
Blue Grass, VA

Dearest Johnnie,
I received your letter. I am always so glad to get them. We are all OK, except Dad. He is sick; he must have the flu. We had the Doctor for him yesterday. If he gets any worse, his heart has been bothering him, and he seems a little better this morning. If he gets worse, we well send you a telegram.
    Looks like if they are going to send you overseas, they will let you come home first. Richard Hull came home. They are sending him to Germany. Maybe it's because he had a wife and baby. I do hope you can get to come home.
    They sawed all this week except yesterday. I got a check from Smith for $390.00 for what they sold him. The Wolf Company hasn't paid for that Cherry and Walnut yet.
    You have been there a month; today seems longer than that. I wrote you a letter this week. Seems like I want to be writing to you all the time. June and Kathleen were here, and Howard Mullenax. Miss Wickline was here Sunday. It was Donald's (Nephew; Joe's son) birthday., We had chicken; I thought about you. I hope you are getting plenty of good things to eat.
    We are having nice weather. Now it is Indian Summer, I guess. I can't think of anything to write. Be good, and God be with you.

    Answer real soon.
    Love, Mother.

Letter #46

October 22, 1950

To: Edith Zinkhan
Frank, WV

From: Sgt John W. Mullenax
408 QM Co 11th Air Div.
Fort Campbell, KY

Dearest Darling,

 How are things by now? I received a letter from you today, but didn't get one yesterday.

 We had inspection today, but it didn't pass so we are working tonight preparing for another one tomorrow. Boy, the boys are sure raising hell. I don't like it too well myself. I think we should get the weekend off.

 Honey, I put in for a pass. I may get it next weekend. Hope so. Darling, I miss being with you, we got along so good, I think we have had some wonderful times together, which I will never forget. No one can take your place. You tried to please me in ever, respect, which I appreciate.

 Darling, I hope I don't have to pull the 21 months and we can start where we left off. Which I mean preparing for the future together. I moved in a room where there is only 3 of us, which makes it better there isn't as much noise and I can sleep. Honey, make sure about the coal, whether you are going to get it right away.

 When you come down to Clarksville, Call, 408 Quarter Master, Fort Campbell and ask for Sgt. Mullenax. Well, Darling I guess there isn't anything to do, but go to bed. I guess you are going out. I lay here and look at your picture and think about you.

  Goodnight, Darling.
  Don't let my love for you be in vain.
  Love,
  Johnnie

Letter #47

October 23, 1950

To: Sgt. John W. Mullenax
408 QM 11th AB Air Div.
Fort Campbell, KY

From: Edith Zinkhan
Frank, WV

Dearest Johnnie,
    How is everything, fine, I hope. I'm fine, but lonesome for you. The trial of Ernest Wimer and Hope was the big event around here yesterday. It lasted all day and Hope won. Everyone knew Wimer didn't have a chance. That was the first time for me and I hope it's the last. Ben was the one that got me in it. He told Hope that I asked Charles to see his blue spot.
    Ben went to the movies last night to see Trigger Jr. and Roy Rogers, but I was tired so I didn't go. I made a custard pie today and have a beef roast in the oven. Would you like to have supper with us? Ha. I sure would love to have you. It will probably be sandwich for Ben gone, I can't keep track of him.
    Hon, excuse this writing, for my right arm is so darn sore. I have a boil just below my elbow and it hurts worse than a toothache. Mom will be here sometime next week, when Floyd and Ruth Collins go, to Huntington. I haven't heard from her since. I wrote and told her they would bring her up.
    Hon, I sure do love you and I hope your love is as strong as mine. I couldn't go out on you, for I love you too much. I'll be the same as you left me and hoping we will have our day, together some day.
    Yes, I meant to tell you Ralph left today for the same camp and said he would see you on Monday or Tuesday. Well, Sweetheart, I'll sign off for now.

With all my love for always,
Love,
Edith (Edy)
Your future.

Letter #48

October 24, 1950

To: Sgt John W. Mullenax
408 QM 11th AB Air Div
Fort Campbell, KY

From: Edith Zinkhan
Frank, WV

Dearest Johnnie,

    Received your sweet letter today and sure glad to hear from you as always. Honey, I didn't mean that I didn't have anything to look forward to in the future. I meant everyday life, Monday is the same as Sunday. I know what you mean, Hon, and what we mean to each other. I guess we are part of each other and you know I love you.

    Yes Hon, Ralph Stone, is there by now. He said he was in your outfit. Boy, Hon you won't be out any too soon for me. Elsie was over for a while and yapped for an hour. Ben took my pen, he just brought it back after sometime. No, hon, don't join the parachute outfit, I do want you come back to us.

    Hon, if you shouldn't come, I wouldn't even want to go on. I know I have Ben, but there won't be too many years he's gone to be on his own. You have just got to come back.

    This darn boil is no better, but it sure is miserable. Everything here is about the same as you left it. I don't go anywhere to find anything out.

    Hon, you know I have faith in you, and I'll trust you, I guess this place and you gone just gets me down sometime It is so hard to put feeling into words. You know what I mean don't you. We have had 3 ½ swell years together Hon. It will soon be 3 years since you gave me my ring, remember.

    Well, sweetheart, I'm running out of news, so I'll sign off for tonight and I love you very much so your Old Battleax will say goodnight and the sweetest dreams, not bad ones. Ha.

    Love for always,
    "Edy"

Letter #49

October 24, 1950

To: Edith Zinkhan
Frank, WV

From: Sgt John W. Mullenax
408 QM Co 11th AB Air div.
Fort Campbell, KY

Dearest "Edy"

    May these few lines find you well and getting along OK. My spirits are high tonight, darling. I just heard over the radio that all Reserves will be out in 4 to 6 months. I hope this is true. I tried for a pass last week and I didn't get it, but had to work. I'm going to put in this week and every week from now on until they give it to me.

    I can see you in court, hope you don't make any enemies. I can't hardly write this. The guys have gone mad since they heard the news. They're running through here banging me on the head, in fact they are almost taking this place apart.

    You tell Ben I said to buckle down on the books or he don't get the watch. It is a nice little watch, it is gold. A boy was broke, and I bought it, the crystal is cracked.

    Darling, I'm going to have to stop, I can't write so I will say what I want to say in a few words.

    I love you darling,
    Love for always,
    Johnnie

Letter #50

October 25, 1950

Tuesday Evening

To Sgt. John W Mullenax
408 QM Co 11th AB Air Div.
Fort Campbell, KY

From: Edith Zinkhan
Frank, WV

Dearest Johnnie,
  Received your letter you wrote Saturday. I sure hope you do get home this weekend. The way the news was today. That you would be out in 6 or 7 months, but that sure sounded good; I was up at 5:30 this morning.
  No, Hon, I didn't go out Saturday night. I have gone to the movies once since you left. That was on Saturday night 2 weeks ago. I just don't want to go anywhere without you. I'm lost. I think there is a certain party around here that would love to see me go out on you. So, they could go over the place and talk about it. But, I have no desire to do so, for I love you too much.
  Hon, I have just now gotten my coal; that worry is over, except paying for it. I am paying 20 down. It's 8.50 a ton. I got 7 ½ tons if the bin will hold that much. They are still putting it in. They finally got the coal in, and you should see Ben; he is black, he calls himself helping.
  Well, Sweetheart, I love you the same as always. Ben is taking a bath. Guess we will go to bed soon as he is finished. So, I'll sign off with all my love for,

  Always, Love "Edy"

  P.S.: *I love you.*

I am now back across the 38th Parallel Line   Julie Mullenax Van Meter

Letter #51

October 25, 1950

To Edith Zinkhan
Frank, WV

From Sgt John W. Mullenax
408 QM Co. 11th AB Air Div.
Fort Campbell, KY

Dearest "Edy"
    How are you and Bennie getting along? I feel a lot better since I heard that the reservist are getting out sooner than the 21 months. The sooner, the better I want to get back with you.
    Ralph was here last night after I wrote you. He sure has the blues.
    Honey, I went to the Captain and asked for a pass. I mean the Captain at the motor pool. He passed it, but it has to be approved by the Company Captain. In fact, I have three Captains over me.
    I think I will stay here until I get discharged or until all the reservists get through here. Darling, this isn't the life for me; I have no time off or very little freedom.
    Ralph told me they weren't taking the draftees very fast. The Order said we would be out as soon as they got enough draftees to relieve us.
    I hope to see you this weekend. Goodnight Battleax. I haven't called you that for a good while, have I?
    Sweet Dreams, and be careful who you dream about.

Love,
Johnnie or (Jackass) Ha.

I am now back across the 38th Parallel Line   Julie Mullenax Van Meter

Letter #52

October 26, 1950

To: Sgt John W. Mullenax
408 QM Co. 11th AB Air Div
Fort Campbell, KY

From: Edith Zinkhan
Frank, WV

Dearest Johnnie,
    Here comes a few lines, another day passed, and what a day for me. I washed and tore the house up, and cleaned all my windows on the outside. I will finish tomorrow, and I'm just about faded out. I'm not used to work. Ha.
    You should see Ben's report card. It's terrible. So, hold the watch until the next 6 weeks' tests. I don't know how to straighten him out. He also got him another dog, a male. I didn't get a letter today, and I sure miss it.
    My arm sure is giving me heck, guess I'll have to go and get it lanced.
    The trial is still in the headlines around here. Wimer's are telling that we were all paid to go and tell lies.
    Ben went to church tonight, and they won't have to go back to school until Monday. I sure would be a happy person if you would get your pass this weekend.
    Hon, that sure was nice coal I got. I built a fire at 6 o'clock this morning, and it was still fire at 6 this evening. Some say it snowed some this morning. It sure was chilly.
    Hon, have you lost any weight? They tell me I have lost since you left, but I can't see it. Not much, anyway. Did you pass your last inspection?
    Oh! Yes, Hon, I got my coal yesterday and today. Mr. Kinsner came down and said he was sorry that they didn't get it.

Well, Sweetheart, I'll say I love you with all my heart and sign off tonight. Good night Sweetheart, and the sweetest dreams.

All my love,
Always
"Edy"

Edith Parsons Zinkhan and Sgt. John W. Mullenax

Letter #53

October 30, 1950

To: Sgt John W. Mullenax
408 QM Co. 11th AB Air Div
Fort Campbell, KY

From: Edith Zinkhan
Frank, WV

Dearest Sweetheart,

    Guess you aren't even there yet. It sure was good to have you home, but I sure hated to see you go, and I sure was ashamed of my house.

    It sure will be good to have you home for good. I'll never will be any good without you.

    Hon, I am sorry I got a little too high Saturday. Maybe the next time, we will dodge our neighbor. How about it. I don't think you enjoyed yourself too much. But I hope you were half as happy as I was.

    I really love you, sweetheart. Ben came over about an hour after you had gone; he wanted to see you before you left. Boy, I sure do miss you, honey. I'll feel better when I know you got there safe.

    Mrs. Lambert got the core out of my arm. It's a lot better.

    Just think, Hon, we had 3 whole nights together, but they sure went fast. I hope our married life don't go that fast. I hope we won't waste all of our time being single. Life is too darn short, don't you think.

    Well, Sweetheart, I'm running out of news. Ben and I are alone again. I'm going to start on my house tomorrow and try and keep it up. I know you saw a difference; at least I can do is take care of what we do have.

    I'll say goodnight and the sweetest dreams. With all my heart, I love you.

    Always, Love
    "Edy"

I am now back across the 38th Parallel Line   Julie Mullenax Van Meter

Letter #54 [5]

October 31, 1950

To Sgt John W. Mullenax
408 QM Co 11th AB Air Div.
Fort Campbell, KY

From: Edith Zinkhan
Frank, WV

Dearest (Jackass)
    I sure hope you made the trip OK.
    I have cleaned the house most of the day. My arm is getting a lot better. Do you still think as much of your Battleax as before, or was I a disappointment to you? I know I looked like heck, but my feelings were all there.
    Boy, I sure hope you get out, hon. Do you think you have the points? The paper stated they were going to let them out that way?
    Hon, I hardly know what to write about. Oh, Yes, Paul Collins is in the hospital in Norfolk, VA, with a kidney stone.
    I am at loose ends again. Oh, yes, everything turned out after you left, I mean the 28th. Everything was waiting for you, Ha. Guess you know what I mean.
    Well, sweetheart, I don't know anything to write about. I'll be waiting for your letter. I'll sign off with all my love for keeps, I hope.

    With all my love,
    Your Battleax

    PS: Guess that's just about all I'm or mean to you, but guess I'll have to admit it. Don't guess I should ever think I could come first with anyone. But, I have given my heart to you. Maybe that's why. You are first with me in

---

[5] *Jackass and Battleax are nicknames they'd given each other*

I am now back across the 38th Parallel Line   Julie Mullenax Van Meter

anything or anyone; I have a different kind of love for Ben, guess you know what I mean.

*   I sent your coat to the cleaners. Everything will be clean when you get out. I put your papers and your court summons in with your discharge. Well, Hon Guess I'll close with that.*

I am now back across the 38th Parallel Line   Julie Mullenax Van Meter

Letter #55

October 31, 1950

To Edith Zinkhan
Frank, WV

From Sgt John Mullenax
408 QM Co. 11th AB Air Div
Fort Campbell, KY

Dearest "Edy"
    Arrived OK, only I was so tired that it wasn't hard to go to sleep. It feels like it was too short of a stay. We tried to do what should have taken a week to do.
    We had to sign a paper about how many points we had. I had 30 points. I don't know what it means as of yet, but I think it means something about whether I have to ship overseas.
    I drove today, and I had a hard time keeping awake. Darling, I sure enjoyed myself and hope I can get home again soon.

    Good night with all my love,
    Johnnie

Letter #56

November 2, 1950

To Edith Zinkhan
Frank, WV

From Sgt John W. Mullenax
408 QM Co 11th AB Air Div
Fort Campbell, KY

Dearest" Edy"
    How are things back there by this time? I have been getting along okay.
    Hon, I haven't sent that $30 yet. I haven't gotten a chance to get the money order. I will send it as soon as I get a chance to get the money order. I am supposed to get paid tomorrow. I have to renew my licenses for my car; they ran out the same day I got back. That will cost another $0.05.
    Hon, I sure enjoyed myself when I was back there to see you. It was so short a time, it seems. I'm in hopes that I will ship out of here next month to a camp closer to you. Darling, I have 30 points I don't know what the amount is to get out yet.
    But I know that many don't ship overseas. Hon, don't worry about the house. I know how you suffered with your arm. I know how well you keep a clean house. Darling, would you send me a box of homemade candy and filled cookies like the ones I like so much I know I'm a lot of trouble, but would you do that for me. I would like to see those pictures, especially the one I snapped of you, ha.
    I haven't seen Ralph Stone since that first time. I don't know what or where he went. Darling, let me know how your arm gets. I think I will try for another pass in two or three weeks. I will close for tonight thinking of you, and it makes me feel better to know then I'll have someone I can trust.
    Love always,
    Johnnie

I am now back across the 38th Parallel Line   Julie Mullenax Van Meter

Letter #57

November 2, 1950

To Sgt. John w. Mullenax
408 QM Co. 11th AB Air Div.
Fort Campbell, KY

From: Edith Zinkhan,
Frank, WV

Dearest Jackass,

    How are things going fine? Fine, I hope. Yes, I'm fine; I don't really know. But I do know, I wish you were here and not in this darn Army.

    Bet you were really tired over your trip. If I don't get a letter pretty soon, I won't know what to write about. There is no use writing our future, for it seems so darn far away. I wouldn't blame you if you did change your mind about an old battleaxe like me. Guess I'm not good enough for you. But I'll take your word and hope you don't let me down in the end.

    Hun, if you think it will cost too much to keep all the junk, I have up when we get married. We will sell it and get something that we can keep up. And we will sell the house and get a farm. I guess I am sitting here daydreaming. I better stop, don't you think.

    I have been figuring on the darn bills. Hon, I'll be clear by April, all but the stove. I have some saved.

    I hope I don't know if I can write anymore or not. Darn, kids just about scared me out of my wit, pounding on the house and then running.

    Never anything else to write about. I'm going to sign off and go to bed.

    I love you, sweetheart, and there will never be anyone else, so say goodnight, sweetheart, and sweet dreams.

    Love,
    Your Battleax.

Letter #58

November 3, 1950

To Edith Zinkhan,
Frank West Virginia

From Sergeant John W Mullenax
408 QM Co. 11th Air Div.
Fort Campbell, KY

Dearest Battleax,
    Received a nice letter from you today. I sure was glad to hear from you and to know that you are well and getting along okay. Hon, what did you mean that you were at loose ends again. I don't quite understand.
    It was payday here, and I think. I am being a good boy. Orders.
    Darling, I think as much of you as ever. We did get a little too much, but it had been so long since we had seen each other. We wanted to do what we could for you to have a good time.
    Darling, I hope we can get together soon and do what we had planned to.
    Darling, it is hard to write about anything. But work. I will stop for tonight.

    With all my love,
    Johnnie

I am now back across the 38th Parallel Line   Julie Mullenax Van Meter

Letter #59

November 4, 1950

To Sgt John W. Mullenax
408 QM Co. 11th A B Air Div.
Fort Campbell, KY

From Edith Zinkhan
Frank, WV

Dearest Johnnie,
    Sure, was glad to hear that you got back okay. I sure hated to see you go. I sure hope you have enough points to get out.
    Hon, mom will be here Sunday. Floyd And Ruth left today for Huntington, Virginia. And if you are there Thanksgiving. I'll come down; maybe you could get a pass. I mean, if you want me to come down. I wrote every day this week except last night. I wrote a letter, but it didn't sound right, so I didn't mail it.
    Hon, I think maybe you know too much about how I love you, for the letter I just got never mentioned it. I was darn happy when you were home. Maybe I did the wrong things or said the wrong things; I don't know. do know if you should let me down, I don't know what I would do.
    Well, sweetheart, I have cleaned the house today, and you should be here, and how I wish you were. I'm sorry it had to be so dirty when you were home.
    This is no news, hon, guess I'll go to the movies tomorrow night, for the darn house is driving me nuts. We had awfully hard rain last night. Well, Jackass, guess I'll sign off with all my love for you always for you always,

    Love, "Edy"
    Your Battleax.

I am now back across the 38th Parallel Line　　　　Julie Mullenax Van Meter

Letter #60

November 4, 1950

To Sergeant John W. Mullenax
408 QM Co 11th AB Air Div.
Fort Campbell, KY

From Edith Zinkhan
Frank, WV

Dearest Johnnie,
　　Received your sweet letter and sure glad to hear from you. I sure hope you don't have to go overseas. Hun, I think if you try to get out, you could have the mill. To get out, if you put it strong enough, your father is too old and not able to do the work, I'd try.
　　I got your cookies made, but, Hun, I don't think they are very good. I also got your candy made. Maybe the next ones will be better. The weekends just about get me down; that is when I miss you the most.
　　Well, Floyd and Ruth got back, but they couldn't find mom, so guess she won't be coming. Hun, do you think you will be able to get another pass soon? I sure would hope so, and if you do have to stay in the darn Army. I'm going back to work somewhere and let Betty keep Ben if she will. What do you think about the idea? Hon, I can't take it without you, and Ben is too hard to handle. Guess it is me and the way I handle him. Ben and I went to the movies last night. It was pretty good.
　　I hope your cookies will be good, and sweetheart, they weren't no trouble. I just wish you were here, and I would love those kinds of troubles. It sure is cold here today. It snowed this morning. It really feels like winter. I sure hate to see it come.
　　Well, sweetheart guess that's just about all the news for this lonesome Sunday night.
　　All for now, with all my love.

　　Love,

I am now back across the 38th Parallel Line   Julie Mullenax Van Meter

"Edy"

P.S.: *My arm is okay now, just hope I don't get any more.*

P.S.S.: *Don't worry about the money.*

Letter #61

November 5, 1950

From: Mrs. H. W. Mullenax
Blue Grass, VA

To: Sgt. John W. Mullenax 33883656
408 QM Co 11th AB Air Div.
Ft. Campbell, KY

Dearest Johnnie,
    Received your letter, sure was glad to hear from you and know you got there OK. I am always uneasy about you when you are on the road, there are so many wrecks.
    Was so glad to have you home. Hope you can come home soon again for a longer stay. Our Golden Anniversary will be the 20th of this month and Thanksgiving will be the 23rd. If you could come then. I had my heart set on having a party, but the condition Mildred is in, I guess we can't have it. Although I would of liked it very much.
    But if you can't come at Christmas to, don't come. Thought maybe you could stay longer at Christmas. I would like for you to come at both dates if you can. I hope you have points enough.
    The War in Korea is not looking good at this time.
    They sawed all last week. Got all the logs at Ted's and Bob's. They are going to haul the lumber to Bartow Tuesday for the Borrer Lumber Co.
    Mona hated it so bad she didn't get to see you. She said, she got up in time to see you pull out. They are bringing Mrs. Hammer home today, she is not much better but, she wants to come home so bad, They, thought they would risk bringing her home.
    Your check from the Government is here. Must I send it to you or keep it till you come home?

    Love,
    Mother

I am now back across the 38th Parallel Line   Julie Mullenax Van Meter

Letter #62

November 6, 1950

To: Edith Zinkhan
Frank, WV

From: Sergeant John Mullenax
408 QM company, 11th AB, Air Div.
Fort Campbell, KY

My Dearest Darling,

 A few lines to remind you that your man loves you and is counting the days to be with you. I'm out of here to be with you for always. Read something about if I change my mind had a thought about changing my mind for you. I realize more each day I am in here, how much you mean to me, and I know that we will have things the way we want them soon. You said something about the cost of the house and things.

 Well, the way you want it, I want it, and we are going to keep things that way. Darling, you said you would be about paid up by April. I want you to save enough to come and see me when I move to another camp. I have your $30 but haven't got a chance to get a money order. A close-up so early. Haven't needed it.

 Went to Nashville last night Grand Ole Opry. I had a car loaded with boys, and we couldn't get in; all the tickets were sold early. I was a little boy. I didn't get drunk. I didn't want to mess up, because want to get back to you. If I get off, I will try to go next Saturday, I don't care a whole lot about it, but if I go, I can say I was there. I only hope that you were here to go with me last night. There was someone missing, and that was you. I don't feel right without you by my side, darling. If we were married, I don't think the want to be near each other would be any greater than now.

 I went to see Ralph Stone and Joe Stanton this morning, and they sure are bitching about the army. I am going to get plenty of sleep today. I don't have to work. They get me up so darn early here and work so late to catch up today. Has your mother arrived yet? If I stay here too long, I want

I am now back across the 38th Parallel Line   Julie Mullenax Van Meter

you to come down. This is the lonesomeness place I have been in. I will close for this time with all my love for my future wife.

    Love for always,
    Johnny

Letter #63

November 6, 1950

To: Sgt. John W. Mullenax
408 QM Co. 11th AB Air Div.
Fort Campbell, KY

From: Edith Zinkhan
Frank, WV

My Dearest Johnnie (Jackass),
    Received your sweet letter. How are things going, Sgt.?
    They are the same as ever around here. When I said I was at loose ends, I mean I'm always at loose ends without you. Don't you understand what I mean? I have too much time on my hands.
    Do you think you will get out soon? I sure hope so. I hope you don't go and get drunk. Hon, for I know you. You said you thought as much of me as ever; I hope you do, hon? I hope I don't get to be a faded picture to you. I couldn't be any other way but true and honest with you, for that's the way I feel.
    Say, Bud, your letters are getting pretty short, guess it is pretty hard to write. If you get tired of me writing so much, just tell me. Hon, there are so many things you don't tell me.
    Well, Sweetheart, I'll sign off for tonight, for there isn't anything to write about. Ben is still the same old Ben. He is getting his lessons. Guess I'll go and vote tomorrow.
    Goodnight, Sweetheart, and sweet dreams. I love you, hon, can't you ever write that. Remember how you used to want to hear those 3 little words. But, like you use to say, don't write them unless you mean them.
    Again Goodnight
    All My Love, Always
    "Edy"

    P.S.: Hope you like your cookies and candy. Did you read the funnies?

I am now back across the 38th Parallel Line   Julie Mullenax Van Meter

Letter #64

November 7, 1950

To: Edith Zinkhan
Frank, WV

From: Sgt. John w. Mullenax
408 QM Co., 11th AB Air Div.
Fort Campbell, KY

Dearest Darling,
    Will write a few lines to let you know that I am getting along OK. Hon, I don't want you to worry about me. I want you to know that you can trust me. I couldn't be a straight shooter than I am.
    Darling, you said something that you were writing about how much you love me. That is what I want to hear. You can't write too much of that. I got plenty of sleep Sunday and Sunday night. That is the most sleep I have gotten since I have been here.
    Darling, I want you to come down to the worst in the world. I hear today, we would be moved to a training company the 15th of this month. I don't know how true this is, but I can't get nothing but a 24-hour pass now.
    I think I will get 6 days to bring my car home when I get my order to ship. The only ones shipping now are the ones from 1 to 20 points. I have 30 points. I can't hardly write this for so many in my room listening to your radio. All of us are keeping up with the war radio news.
    Hon, I sure would like to be back there with you, my one and only desire.
    How is Bennie getting along? Tell him I said to be a good boy. Tell him I said it won't be too long until Santa Clause comes.
    How are your neighbors getting along? Well, honey, I will be looking for your letters. Oh, don't forget to send me the box and don't forget me in your dreams.

    Goodnight Darling,
    Love for always,
    Johnnie

Letter #65

November 6, 1950

To: Edith Zinkhan
Frank, WV

From: Sgt. John W. Mullenax
408 QM Co., 11th AB, Air Div.
Fort Campbell, KY

To My Darling "Edy"

    May these few lines find you well and getting along OK. I am still in the same place, but I won't be here long. I just heard today we would ship out in the next 5 days, between now and the 15th of this month. I don't know, but I think we will ship to a training company and take this training before we ship to another camp. Hon, I never heard from you for the last couple of days; I hope you aren't sick. I haven't heard from mother since I have been back, either.

    The boys crowded in this room and talked about whether or not they had enough points. We have some men that only have 6 or 7 points. The ones that are shipping over are from 1 to 20 points.

    We can't get 3-day passes anymore. I was going to try to come home to see you this weekend. Darling, I love you and never will stop loving you. I hope you realize that I do. Darling, I like to hear you say you love me too.

    The time we have spent together, I will never regret, and hope we can continue that. Hon, I have enclosed your money order, and thanks a lot. I can't pay you with a kiss but hope I only could.

    Have you gotten the pictures back? I bet they are something?

    Did you ever get that mix master? I want you to get on them about it. I guess mother thinks she will never get it. Honey, this Army isn't for me. I would like to be back there with you and share our love together.

    How is Bennie getting along? Tell him I don't want to hear the things like I heard when I came back about his grades and not listening to you. If he doesn't listen to you, Hon, take the watch away from him until he does listen to you.

Well, Darling, I will have to shave yet tonight. I don't have time in the morning. Hon, I love you so much. I didn't know I could feel this way about anyone but you. You are the one and only one for me, and in my mind, you feel the same about me.

Good night Darling,

Love for always,
Love,
Johnnie.

I am now back across the 38th Parallel Line Julie Mullenax Van Meter

Letter #66

November 8, 1950
Wed night

To Edith Zinkhan
Frank, WV

From Sgt. John W. Mullenax
408 QM Co., 11th AB Air Div.
Fort Campbell, KY

Dearest "Edy"

 I have some news tonight about the shipment. I will be shipped out of this company between now and the 15th of this month. I don't know whether I will be shipped to a training company to take basic training or I may be shipped to a camp from this company.
 I did hear this; the ones that have less than 20 points will have to get rid of their cars. But the ones that have over 20 points may keep their cars. That sounds good; I think it means I will get a delay in route of 5 or 6 days to the next camp. Which means we will be together.
 Darling, don't do anything about getting a job until I get to the next camp. I don't like to be bossy. But, darling, won't you do this for me. I want you near me, and we would be in a better position to be together.
 Here, I don't know what is going on, and I am subject to ship out at any time. That is why I haven't had you down here with me. Don't feel harsh at me darling, you should realize by this time that I want you near me.
 I sent your money yesterday, and thanks a million. Hon, about the passes, I can't get any more passes because I am subject to move at any time.
 Honey, if I could only get out of this army and back with you. I don't know anything I would like better. There is one thing sure they are only shipping the ones with 20 points and under overseas. I have enough to stay on this side.

I am now back across the 38th Parallel Line　　　　　　Julie Mullenax Van Meter

    Honey, I have your picture here on the shelf near my bunk. I don't have a table. I look at that picture, and all I can think is that woman has my heart and how much I love her and the yearning to be with her.
    Hon, I will close tonight and look forward to a letter from you tomorrow and honey, keep your hopes up, we will win someday.

    Goodnight and sweet dreams, Darling, for I love you more than you will ever know.

    With all my love for always,
    Love,
    Johnnie

Letter #67

November 10, 1950

To: Edith Zinkhan
Frank, WV

From: Sgt John W. Mullenax
408 Q M, Co, 11th AB Air Div
Fort Campbell, KY

Dearest "Edy"

Hope these few lines find you well and getting along OK. I'm getting along OK, but I know a lot of things I like to do than this. I received the box today, and "oh" boy, were they good. The cookies and candy. All my buddies sure jumped in. They said she sure is a good cook. I told them I had known for several years.

Darling, I don't know how long I will be in the army, but I hope it isn't too long. Darling, I hope I get to a camp where I know that I will be stationed for a while. When I do, you are going to be there with me. Is it OK with you? I want you near me, honey.

Darling, I received a letter from mother, and she said her and Dad's Golden Wedding Anniversary is the twentieth of this month. (November). Hon, I don't know what to get them. Can you help me out on this? I would appreciate it very much.

How is Ben by this time? I guess he still wants his way.

Hon, don't stop writing those three little words, and I mean those three little words. They are big to me, meaning a lot to me. Hon, when you get time, you can send me another box.

Darling, I will close for tonight, knowing that I have someone to look forward to. Honey, I love you and don't forget to tell me.

Love for always,
Love, Johnnie

I am now back across the 38th Parallel Line   Julie Mullenax Van Meter

Letter #68

November 11, 1950

To Sgt. John W. Mullenax
408 QM Co, 11th AB Air Div.
Fort Campbell, KY

From: Edith Zinkhan
Frank, WV

My Dearest Johnnie,
    Here come a few lines, and hoping they will find you the best. We are OK, only I sure would love to see you. I have cleaned house again today. I wish you were coming. Guess, I'll have to keep on wishing, don't you think? I didn't receive a letter today. Hope you haven't shipped out.
    Hope you have received some mail from me by now? You should have, hon, for I, have written every day. It has turned cold the last couple of days. Have you heard from your mother yet? Hon, I know she was hurt because you didn't stop over there first. Guess I'll go to the movies tomorrow night. And now I wish you were here to go with me. Hon, please write and tell me if you have to be shipped overseas. Don't keep it from me.
    I saw about the mixer, and he has it fixed. All but painting it, So, when I get it, I'll take it up to your Aunt Gertrude, and maybe some of them will come and get it. I also sent the pictures away, and soon as I get them, I'll send them to you. I also ordered your razor. I'll wait and give it to you. If that's alright, I sure hope we can be together for Christmas. Well, soldier, I love you very much, and there will never be anyone but you. So, hurry up and come home, so we can be happy together. But of course, we have to wait until Uncle Sam says so. Hon, this sure is a mixed world.
    Sweetheart, there isn't any news, so I'll sign off with all my love, and I do love you squirt and will for always.

    Love, Your Battleaxe

Letter #69

November 14, 1950
Monday night

To Edith Zinkhan
Frank, WV

From Sgt John W. Mullenax
408 QM Co., 11th AB Air Div.
Fort Campbell, KY

Dearest "Edy"

    Will write a few lines to let you know that I arrived OK. Darling, I had a wonderful time; what time was I home? It's a shame so short a time.

    Hon, I have moved tonight to a processing area. I will be here a couple days and then ship to a training company to take the three weeks training. I heard today that the point system is still in. I am glad about that. Darling, use the old address. Don't change until I get to the training company.

    I am sure tired tonight. I got in last night from 10 till 3. Darling, I hated I rushed off like I did, but the time slipped up on me.

    I am still with my buddies. Well, hon, I will have to get some rest, so I will say good night, sweetheart, and sweet dreams. I love you, darling; maybe I don't say it often, but I sure mean it, and I want you to believe me and trust me.

    Love for always,
    Johnnie

Letter #70

November 19, 1950
Thursday eve.

To: Sgt John W. Mullenax
408 QM Co., 11th AB Air Div.
Fort Campbell, KY

From: Edith Zinkhan
Frank, WV

Dearest Johnnie,
    I received your letter and sweetheart, I have written every day, but one since you left and I sure know how you feel when you don't get any. I'm not sick, only headaches, and I think that's from my eyes.
    Hon, I sure do love you. I feel like your part of me since you have gone. I just can't get myself adjusted and don't think I ever will until you get back; I sure hope you don't have to go overseas and hope you're shipped closer home.
    Hon, you didn't have to worry about that money. I will never be able to repay you for what you have done and given me. I sure hope you get your cookies. OK, and hope they are good. The candy didn't turn out good.
    Hon, Do you know what date Sunday is? It will be 3 years since you gave me my ring, and it's been 3 ½ years since we met. I haven't seen anyone from over your home. I don't think they approved of me. But it's you I love, so guess that is all that matters.
    Ben is trying a little harder, I think. One night since you have gone into the Army, he said, I sure hope Johnnie gets out soon. I asked him why. He said I was mean to him when you weren't here. I guess I'm a little hard to get along with since you aren't here.
    I just finished supper. I have been making myself eat. Ben is getting his lessons.
    Well, sweetheart, guess that's about all I know to write. I could write and respond by writing how much I love you and how much I care.
    I'll sign off, Hun

Love for always,
"Edy"

Letter #71

November 24, 1950

To: Edith Zinkhan
Frank, WV

From: Sgt John W. Mullenax
Co. B 767 B. N. 11th AB Air Div.
Fort Campbell, KY

Dearest, "Edy"
    Arrived back to camp OK. I got here at 7:00 pm in the evening. I had a good plane ride; it wasn't rough. We had 2 stops between Nashville, TN., and Roanoke, VA. One was Knoxville, TN, and at Kingsport, TN. It cost me $28.40, and I got a free meal included. I had a headache last night; I went to bed and slept it off. I didn't have much to do today. I am going to transfer to another Company because I missed a few days to take my car home.
    I don't know when I will move, but it will be after Thanksgiving Day. I hope I hate to move now. I made it in good time to catch a plane, but it sure was slick over the mountain. I didn't know whether I would make it or not without chains, but I got along OK
    Mother, Aunt Gertie, Mona, Joe, and Milton went with me to Roanoke.
    Hon, I had a nice time at home, more so than the other 2 times. If it had been a pretty Monday when we went hunting, we would have had a nice day, but it had to be raining. I have been using my razor, and it sure works fine. Darling, they just told me we would take four weeks of training now instead of three. In that case, I will be here for Christmas, "I hope." Write me at this address, my buddies will hold my mail for me. Darling, I guess I will get Turkey tomorrow, but it won't be anything as you cook. I told them at home to bring that deer head to you, so you put it where you want. I will close for this time. I love you, Darling, and know that I won't change.

    With all my Love for Always,
    Love,
    Johnnie (Jackass)

Letter #72

November 24, 1950

To: Edith Zinkhan
Frank, West Virginia

From: Sgt John W. Mullenax
Co B 76th TK BN
11 HB Div.
Fort Campbell, KY

Dear "Edy"

  This is supposed to be Thanksgiving Day, but any holiday in the Army doesn't seem the same as being back there with you. I had a very good meal today, but nothing tastes as good as my little cook.

  Darling, I sure appreciate your cooking. It is a pleasure to set down at your table for a meal. Darling, it looks as though I will be here in this camp for Christmas. I have to ship to another Company because I missed a few days' training and they are going to give us four weeks' training.

  I am going to try and get ten days for Christmas. I went to a movie tonight with 3 other guys. I didn't like it too well. The name was Wild Fury.

  The war news sounds good, as if it may be over in a matter of a few days. I hope I think we will get out sooner it is, don't you think. You know that Bruister Girl you were talking to? Well, Hise got a box of cherry candy from her, and I helped him eat it.

  Darling, it is true about the points; they hold the men out that have 20 points and more at the next camp.

  So don't you worry about me, honey; I will get out before too long, and then you and I can accomplish what we wanted to. Honey, I sure enjoyed you being with me hunting. If you wouldn't have gone, I wouldn't have gone either.

  Darling, Dad, and Mother likes you a lot, it seems. Like Dad wants to be around you. Didn't you notice it while we were hunting? Hon, these days that we spend apart make us realize what we mean to each other and that we should always be together.

Darling, I will say good night for tonight, and I love you so much. If we were married, I couldn't love you anymore. I hope that we can be partners in the near future.

Love for Always,
Johnnie.

Letter #73

November 25, 1950

To: Edith Zinkhan
Frank, WV

From: Sgt John W. Mullenax
Co. B 76 TK. BN.
11th AB Air Div.
Fort Campbell, KY

Dearest "Edy"

How are you by this time? Fine, I hope. Has it been cold back there? It has been about zero here, and I have been out in it training. It sure is hard to take this kind of weather. It is so damn cold in this barracks that I have to stay in bed to keep warm. I thought I would move to another Company, but they gave me my field equipment. So, I guess I will stay here.

From the rumors I have heard, I will get ten days for Christmas. I hope so, Honey. I sure want to spend the holidays with you. I would like to have Mom and Dad over one day if you don't mind. Darling, I sure will appreciate that box. When I get it because this food is lousy here. By the way, Ralph Stone was up to see me this afternoon. He was pulled off of the shipment detail. He said that his wife went home, so go up and tell her that he is still here. He sure was tickled that he could stay.

You know that picture I wanted to bring. Send it to me and send one when I'm over home.

Darling, I haven't been writing too regular, I hope you don't get sore at me, but they have been pushing me around; I've had to scrub barracks last night and had an inspection at 9 o'clock. You can imagine how cold and damp it was in here, and we stood another inspection this morning.

Darling, I love you and can't help to think about you by yourself back there, and what it is like to take care of Ben, because he is getting so big. Tell him I said to be a good boy and to listen to you because I want to come home without hearing he wasn't listening and not getting his lessons.

Tell him I said I was sorry I didn't get to take him hunting. But that we will have lots of time to go hunting when I get out.

Darling, I miss you lots and hope I don't have to stay away from you too long, for I love you, and I know that you love me too, which makes it so much harder.

So good night, Darling. Sweet Dreams, and I love you.

Love, Johnnie

Letter #74

November 26, 1950

To: Ben Zinkhan
Frank, WV

From: Sgt John W. Mullenax
Co. B. 76 TK BN. 11th A.B. Div.
Fort Campbell, KY

Hi Ben,
    How is my boy getting along? I presume you are getting your lessons every night and listen to what your mother tells you to do.
    Don't forget, buddy, it isn't too long until Santa Clause will be coming, and I know you have something in mind for him to bring you, so it pays to be a good boy. I hated I didn't get to take you with me hunting, but there will be other times. We can go when I get out of this Army. Of course, there are a lot of things you must learn about the handling of a gun before you can carry one yourself.
    Well, Ben, I have a lot of work to do this afternoon. But, I am trying to get a pass to see you Christmas, so be a good boy and listen to mother.

    Love,
    Johnnie

Letter #75

November 26, 1950

To: Edith Zinkhan
Frank, WV

From: Sgt John W. Mullenax
Co. B, 76 TK. Bn. 11th A.B. Div.
Fort Campbell, KY

Dearest "Edy"

Will try and write a few lines today. Although I haven't received but one letter from you since I have been back.

We don't have any mail call today, so I will have to wait until tomorrow to hear from you. It is so darn cold down here. I haven't been out of the building only to eat, and still, I don't have much to eat. I sure know how to appreciate your cooking and the love you share with me.

I sure will be glad to get the box you are sending me. Honey, I want to get this training over with. Then I know I won't have to go through it again, and I will have a better chance of getting a furlough at Christmas.

I can't send my clothes to the laundry, so I will have to wash some this afternoon. I am still in the same Company, so I guess I will go ahead and finish my training here in the Company. I have 2 1/2 weeks of it yet.

I had to sign a paper when I got my call, and when I reposted, they said I didn't have 14 days, and I would get a 10-day furlough, but I don't know how true this is; it is hearsay.

Darling, if you have a chance, get us some drink for the holidays in case I get home. Hise and I are still together. I hope we get to come home together. I hate to travel by myself. Darling, I am glad I don't have my car with me because the roads are nothing but a sheet of ice.

Darling, I am proud of my razor. It works fine; what I mean is it won't take long to shave. Honey, it would surprise you at the men in this Company that last less than 20 points. That is what pulled Ralph Stone off the shipment. I think, we don't know why, but that must be the reason for pulling him off. He sure had a pleasant smile when he came to tell me. Honey, have you decided what to get Ben for Christmas. I know you want to go shopping and I hope you get away. Try and get something for Dad and Mom, and I will give you the money when I get home, if you don't mind.

How are you making out with your neighbors? I guess it is the same old thing as usual. What did they say this time about us? I guess they made up something. Well, Darling, I will have to close this time. Hoping to hear from you. How are you getting along back there, and all the news? Tell all I said Hello. Darling, I love you and hope you believe me because I mean it.

So, I will close with all my Love for always,

Love,
Johnnie.

Letter #76

November 28, 1950

To: Edith Zinkhan
Frank, WV

From: Sgt John W. Mullenax
Co B. 76 TK BN, 11th H B. Div.
Fort Campbell, KY

Dearest "Edy"

How is my Darling getting along, by this time? I hope these few lines find you well and all things are going along OK. Honey, I haven't heard from you for 2 days. It may be because of the snowstorm. I sure like to hear from you, darling.

Honey, it keeps my spirits up to hear from you and to know I have a lovely woman like you waiting for me. I know it is lonesome for you because of the time we spent together, and we were so close to each other. I don't think a married couple could have been hit any harder by parting than we were.

We are bound to have a break sooner or later, so keep your spirits up. I have mine.

I have been out at the firing range today. It was rather cold, lying on the ground and firing the rifle. My ears are still ringing from the rifle fire. I can't hear well.

Honey, the days are slipping by and getting close to Christmas. I hope I can spend the Christmas Holidays with you because it won't be Christmas without you. Do you know whether Ralph has shipped out yet? I haven't seen him for a while. Are they drafting many boys back there? The sooner they get them, then the sooner I get out.

Tell Ben Hello, and not to forget what I told him. Because it isn't long to Christmas. You haven't sent me the pictures yet. I am anxious to see them.

I am now back across the 38th Parallel Line   Julie Mullenax Van Meter

Well, honey, I must close and get cleaned up because our lights go out before too long.

Sweet dreams Darling; I love you and hope we may never be parted again, and may our love never perish.

Love for always,
Johnnie,
Your future hubby.

I am now back across the 38th Parallel Line   Julie Mullenax Van Meter

Letter #77

November 28, 1950

To: Edith Zinkhan
Frank, WV

From: Sgt John W. Mullenax
Co. B, 76 TK BN 11th A.B. Div.
Fort Campbell, KY

Dearest "Edy"
 Received your letter today and am so glad to hear from you and to know that you are getting along OK. I sure was glad to receive the photos. The boys got a laugh out of the picture of you when I told them how it happened. Honey, I can't understand why you haven't heard from me. I have been writing.
 This is the first day I have heard from you in 4 days. I have been out in the field today, firing the rifle and grenades. I eat out there, and I got pretty darn cold. I guess I go out on barracks or sleep in pup tents this Thursday, for as I know. Use this address because I am going to stay with this Company.
 Well, honey, I will have to close for tonight. OK, by the way, I would like a little daughter just like you.

Good night Darling.

I love you and only you.
Love, Johnnie.

Letter #78

December 4, 1950

To: Edith Zinkhan
Frank, WV

From: Sgt John W. Mullenax
Co. B, 76 TK BN; 11th A B Div.
Fort Campbell, KY

Dearest "Edy"

    I am so glad to hear from you and to know that you are getting along OK. I have been receiving your letters regular. I got the box today. Honey, I sure appreciate it. Everything was so good that I had eaten too much. I gave my buddies some, and they said she sure is a good cook. My remark was I had known it for quite some time. Honey, you are the last cook I know of. It is a pleasure to eat something you prepare.

    Hon, is Adam and Daisy getting along? Is he still working at his old job? I guess he is awaiting this Army out.

    Honey, I got a letter from mom. She said they got back from Roanoke, OK. She also said she appreciated the things that you and I got them and that she put the cake in the deep freeze, so I guess you and I will have to go over and help eat it at Christmas.

    Hon, how has Ben been acting since I left? Has he been a good boy? Tell him when I come home. I don't want to hear otherwise. I bet he is talking about what he wants for Christmas. Honey, I will get a bottle when I get home, so don't worry about that. I think that they will have the roads open before Christmas, so you can go to Elkins on the bus, Darling. I love you and hope you have realized it by now. Hon, I am going to a movie tonight. I am getting so tired of this bunk and all that goes with this Army.

    I have to go back to pup tents Monday morning and stay until Wednesday. I am going to take some of the cookies with me if I can. Or if there is any left by then.

I am now back across the 38th Parallel Line   Julie Mullenax Van Meter

    Hise just came down and wanted to go to the movies, so, darling, I will close for this time with all my love, and may the future be brighter for us.

    Good night Darling, love for always,

    Love,
    Johnnie

Letter #79

December 4, 1950

To: Edith Zinkhan
Frank, WV

From: Sgt. John W. Mullenax
Co B, 76 TK BN, 11th AB. Air Div
Fort Campbell, KY

Dearest Darling,

    Received your letter and box of candy today. Thanks a lot, it is the best candy I have eaten so does my buddies. The guy on the next bunk got a box of candy from his wife. All I can say is that she doesn't know how to make candy. But I didn't tell him so.

    I just got back from sleeping in the pup tent last night. I darn near froze. I have to go back out Monday and stay until Wednesday. We hiked out about ten miles. I got blisters on the bottom of my feet. I thought I wouldn't make it, but I got there. Otherwise, I didn't mind it too much.

    I got the pictures you sent me. I believe I told you in one of my other letters. The weather is better down here today. It is the warmest in this barracks than it has been since I have been down here.

    Honey, I am sure surprised about what you said about Daisy Lee.

    Darling, as far as I know, I will finish my training on the 18th of December. Hope I get shipped to a camp not too far away from you. The men got paid today, but I didn't. My name wasn't on the payroll. So they wouldn't pay. I went to the 1st Sgt. He is going to check on the matter.

    Well, Honey, I am going to get some sleep tonight. I didn't get but a little last night. I also have an inspection tomorrow.

    Goodnight, Darling, I love you.
    Love for always, Johnnie

Letter #80

December 6, 1950

To: Edith Zinkhan
Franklin, WV 26807

From: Sgt John W. Mullenax
CO. B 76 TK. BN. 11th AB Air Div
Fort Campbell, KY

Dearest "Edy"
    Received two lovely letters from you today. I sure was glad to hear from you and to know that you are getting along OK. I have been getting along OK so far. I didn't have to go to sleep in the pup tent today, so I will leave in the morning. I won't be able to write for a couple days. Out there, I have the fruit cake left. I have a hard time to keep the boys out of it. I want to take it with me tomorrow, if I can, to eat out there. I received the boxes, and I sure do appreciate the things you sent.
    Honey, will you send another book of stamps? I am out. I will owe you a lot of money when I get home. I will pay you when I get home for Christmas.
    Honey, I know I made a mistake when I signed to the reserves, but we were misled. They told us that we wouldn't be called unless war was declared. Which it hasn't, and not only that, the organized active reserves were called, and not only that, they get paid. It all depends on the war in Korea. How long I will be in here.
    Only I hope it won't be too long before I get out. When I ship out from here, I will know more about what I will do and whether I will stay in the states by taking my training about the last of all the reserves. There are a whole lot of Reserves ahead of me to go somewhere when I get this week in here. I will have the worst of my training finished then. Honey, I sure would like to be back there with you.

I am now back across the 38th Parallel Line   Julie Mullenax Van Meter

    Well, Honey, I will have to make up my pack tonight and get ready to move out in the morning. We get up at 4:30 am, and I don't have too much time in the morn.

    Darling, I still love you. I think we know what we want as long as we have gone together; that is, we want each other.

    Good night Darling,
    Sweet dreams.
    Love,
    Johnnie

I am now back across the 38th Parallel Line   Julie Mullenax Van Meter

Letter #81

To: Edith Zinkhan
Frank, WV

From: Sgt John W. Mullenax
Co B. 76 TK BNI, 11th AB Air Div.
Fort Campbell, KY

Dearest "Edy"

    Received a letter from you today that was written Sunday. Hon, I didn't have to stay out all night. We came in at about 11 o'clock. It turned sleeting and raining. We were covered with mud. It was a mess to clean up. We will take a ten-mile hike tonight. We start at about 7, and it sure is raining here. In fact, it has been raining all night.

    Honey, I got your packages. The cookies and candy were really good. The fruit cake, I took it out with me, so I didn't stay, but when I got back, it sure tasted good. How long does Aunt Gertie want you to work?

    Honey, why are you losing so much weight? You aren't taking those damn pills, I hope. I am glad you have gotten glasses that fit your eyes. I am glad that Daisy came up to visit you; I know how lonesome you must be. If you work for Aunt Gertie, that will take some of your time. How will you get up there and back when Ben gets out of school, or will you have him come up there. Hon, I ask you for another book of stamps in one of my letters, but you may not have gotten it. Hon, I am looking forward to seeing you at Christmas. The way it looks, I may be home, and I will have my training through. Find out whether Ralph got his furlough from here or at his next camp. There is a group of men going to fill this Company up soon, so that means we will have got out for them. They are permanent men for this Company. Well, Darling, it is about time for me to start on the hike, so I will close for this time.

      With all my love,
      Love for always,
      Johnnie

Letter #82 [6]

December 9th 1950

To: Edith Zinkhan
Frank, WV

From: Sgt John W. Mullenax
Co B 76 Tk, BM, 11th A.B. Air Div
Fort Campbell, KY

Dearest "Edy"
    I received your letter yesterday and am so glad to hear from you. Hon, I am writing this letter this morning. I was so darn tired and cold last night, and it was so late. I went through the infiltration course yesterday and last night.
    I got the photo of Dad and Mom. Thanks a lot. How is the work up at Aunt Gerties? Don't work too hard; I don't want you washed out when I get home. Hun, the hardest part of my training is finished.
    They got the draftees in here yesterday, and we cannot talk to them or even go over to their hut. Hun, I love you, and I will never change my mind. Hon, I hope to be home for Christmas. Well, squirt, I will have to close. We are going to have an inspection this morning.

    Hope to hear from you today. Darling, I love you always.
    Love, Johnnie

---

[6] *The Infiltration course was one of the final tests of basic training. It involves crawling through an obstacle course under barbed wire and across varied terrain. Often live ammo was shot over the trainees backs and small explosions triggered to mimic the feeling of real combat.*

I am now back across the 38th Parallel Line   Julie Mullenax Van Meter

Letter #83

December 9, 1950

To: Edith Zinkhan
Frank, WV

From: Sgt John W. Mullenax
Co B 76 TK BM. 11th AB Air Div.
Fort Campbell, KY

Dearest "Edy"

Will write you a letter; I know you are getting tired of those little notes I have been writing. Hon, there isn't a whole lot to write about, only the Army. I know you don't want to take this because I am. There is one thing important to write about, and it means everything to me, and that is I love you. I have never cared so much for anyone as much as I do you. I think you and I have spent lovely times together. I hope we can spend some more of these kinds of moments together.

Honey, all I can say is that I have tried to make you enjoy yourself what time I have been with you. I know I have enjoyed myself with you in a way that words can't describe. How much I have enjoyed myself. Honey, I often think about the time when we were at the breakfast table, and we hadn't gone together too long, and you said, "It was only me," Ha, Ha. That isn't nice to think about, but I have to laugh to myself. Honey, I have been borrowing writing paper; I haven't gotten over to the PX to get any, so excuse the paper and writing.

By the way, I didn't get a letter today. Hon, one of the boys told me that the mail clerk didn't take the letters out, so that may be why you haven't been hearing from me. I have finished my training today, and now I will be processing until I ship out, when I don't know.

I sure have a lot of dirty clothes to wash. I will spend my weekend washing clothes.

How is the job going along? I guess they are rushed through deer season. Honey, I got the picture of Dad and Mom. It was good. I got paid today. I will send you the $20.00 as soon as I can get it over to the post office. Hon, I got a letter from Mom saying they haven't run the mill since deer season, which I don't like. All I hope is they don't break the Company up because we have to have that to get a start.

    Honey, how are you getting along with the bills. Have you about got them paid up? I would like to see you on top. I know it has been a lot of responsibility for you.

    Darling, I will have to give you credit. You have done more than some men have. Battleaxe, I love you more than you will ever know because I don't know how to put it in words. Actions speak louder than words.

    How is Ben getting along? I received a Christmas Card from him. I haven't sent any as yet. I will say good night to my future wife and hope she doesn't forget her future husband.

    Love,
    Johnnie

I am now back across the 38th Parallel Line   Julie Mullenax Van Meter

Letter #84

December 11, 1950

To: Edith Zinkhan
Frank, WV

From: Sgt. John W. Mullenax
Co B 76 TK BN, 11th AB Air Div.
Fort Campbell, KY

Dearest Darling,
　　Will write you a few lines to let you know I am getting along OK. It is so darn cold here in our hut. I can hardly write. I was over to the orderly room just now and told them off about no heat in here. Two of my buddies walked down to the movies last night. It was a Western. All through the movie, I would think when I was back there, you were always beside me when I went to the movies.
　　The latest news rumor I have heard is we are shipping out on the 14th of the month. I hope it is up north, somewhere close to you. I often look at the draftees and see how I acted when I first came into the Army. Most of them are young boys here.
　　Honey, I am looking forward to being with you at Christmas. I want to spend it with you and Ben like the last time I was home. Not too many friends. I know you will have a fine Christmas dinner. I know I will appreciate the dinner, but being with you is what I want most.
　　Hon, I don't want you to overwork yourself working up to Aunt Gerties and doing the work at home too. Hon, are you getting back and forth up there? Darling, I love you; I don't think I can tell you how much I love you because I don't know how to put it on paper.
　　How is the weather there? It is so darn cold here. My fingers and toes are so cold. I am going over to the service club to get warm. Darling, I will close at this time.
　　With all my love. I know that someday we will be able to share our love as we would like to.
　　Love for always,
　　Love, Johnnie.

I am now back across the 38th Parallel Line   Julie Mullenax Van Meter

Letter #85

December 11, 1950

To: Edith Zinkhan
Frank, WV

From: Sgt John W. Mullenax
Fort Campbell, KY

Dearest Darling,
    Received two letters from you today. I sure was glad to hear from you. I can't understand why you haven't been getting my letters. I received the stamps today; thanks a lot. I also received a letter from Mother today. She said that she was sending you some meat. I am glad that you got it.
    Honey, I got your Christmas present today. I'm sure having a hard time trying to wrap it. I can't get any wrapping paper. So, I will wrap it the best I can, although it doesn't look so good. I only hope you like what is inside. I don't know when I will get it mailed, but I will mail it as soon as I can get away to do it. They keep a close watch on us here. I finished my training today. I sure am glad about that. They didn't teach me anything I didn't know.
    Darling, I look to being shipped from this camp at the last of this week. But I'm not sure about that and don't worry about it. Keep the letters coming. I will let you know when there is anything definite.
    Darling, I want to know what the reason is for you losing so much weight. If you aren't eating, I want you to start up right now. Because that will hurt your health.
    Mother said they took Meade back to the hospital Monday. It looks like he won't get well.
    "Hon," I couldn't get Ben anything down here. They don't have anything on the post. I may be able to get something on my way home. Darling, I will close with my love, and I will always love you no matter what happens.
    Love for Always,
    Love, Johnnie (Jackass)

Letter #86

To: Edith Zinkhan
Frank, WV

From: Sgt John W Mullenax
Co B 76 TK BN; 11th A. B. Div.
Fort Campbell, KY

Dearest "Edy"

    Received a sweet letter from you today. I'm so glad to hear from you and to know that you are well and getting along OK. I am getting along OK. I have been waiting to get transferred to a shipping port at the other end of the camp. So don't worry about it, darling. I still think I will get a furlough if I don't get it here. I will get it at another camp.

    From what I hear, I will move down there tomorrow, and when I ship out, I don't know. Some of the boys are getting their furloughs from here. I think Hise is getting his furlough from here.

    Darling, I still have your present here. I haven't been able to get to the post office. It is way across the camp, but I will get it to you somehow.

    Darling, I love you and sure would like to be back there with you. by being away from each other, I realize what you mean to me more than ever. I am glad you get some meat. It will help you out that much.

    How do you like your job up at Aunt Gerties? Do you have to work hard up there? I haven't been doing anything for the last couple of days, only some shots, but they keep up close.

    Hon, I would appreciate a box from you anytime. You are the best cook I know of. I have eaten some of the things some of the boys have gotten from home, and none of it will compare with yours. Just to prove it, all of the boys want to know when you are sending a box.

    By the way, don't send more boxes here.

    Well, Hon, I will close with all my love, and may my love for you not be in vain.

    Love,
    Johnnie

Letter #87

December 15, 1950

To: Edith Zinkhan
Frank, WV

From: John W. Mullenax
Co. B 76 TK BN
Fort Campbell, KY

Dearest Darling,

 How is my darling future wife by now? I received a lovely letter today and a Christmas card. Darling, it does me so much good to hear from you and to know you are getting along OK. I sit here and think of the lovely times we had together; all I want in this world is to have those times back and have you.

 Darling, I know we haven't done what is right, but I don't care because my intentions are good, and I love you not as a girlfriend or a fiancé even greater than that.

 Hon, I got to the post office to mail your package today. I hope you like it.

 Hon, I don't think I will get to spend Christmas with you. Don't feel too bad about it. Darling, there are guys here that have been here as long as I, and they haven't been home.

 Darling, I am shipping out Monday the 18th to California. If I get stationed down there, I'm sending home for you. I will send what money I have for you to come on.

 I can't tell you how much I would love to have you with me.

 I guess you have everything for Christmas prepared. Hon, I would love to eat that Christmas dinner of yours.

 I saw the band playing for the boys that shipped out today. I guess he was; I mean, Ralph Stone was in the bunch. I guess they will play for us when we ship out Monday.

I am now back across the 38th Parallel Line　　　　　Julie Mullenax Van Meter

    How has Ben been acting lately? Good I hope. Tell him I wish him a Merry Christmas and hope Santa Clause brings him nice things.
    Darling, all I ask is to believe in me, for I will be back to you, and we can start out where we left off. Well, I will have to close tonight with all my love to the sweetest girl I know, with Sweet Dreams to my future wife, and a Merry Christmas and a Happy New Year.

    Love for Always,
    Love, your future hubby
    Johnnie.

Letter #88 [7]

December 17, 1950

To: Edith Zinkhan
Frank, WV

From: Sgt. John W. Mullenax
Co B 76 TK BN
Fort Campbell KY

Dearest "Edy"
    Received a lovely letter from you, which I am always looking forward to. Hon, I pull out in the afternoon for CAMP STONEMAN, CA. tomorrow. I don't know whether I will ship overseas or not, but I still have hopes.
    Darling, I tried to call you this afternoon but couldn't get you. I guess you should stop writing until you hear from me and then your letters will come to me sooner. Otherwise, it will be a long time before I get them.
    Hon, it doesn't seem possible that I am to be away from you this Christmas. Although I will be thinking of you. Hon, I got your Christmas package mailed. It was the best I could get in this camp. They don't have much to pick from. It is to the dearest and loveliest woman I know and the one and only that I love.
    I couldn't find anything for a boy here; I may be able to get something at another camp.
    Darling, if I do get stationed somewhere in the states, I want you to come where I am at. If I do get stationed somewhere, I will let you know, and we can make arrangements then.
    Hon, if you had a driver's license, you could have my car to use. Only I wouldn't want you to take it on slick roads.
    That present you got for me I like it better every day. It saves time and a lot of trouble. Well, I guess I will have to close this time by telling all Merry Christmas.
    Love for always,
    Love, Johnnie.

---

[7] *Camp Stoneman was named after George Stoneman, a cavalry commander the American Civil War and a Governor for CA. It opened May 28, 1942 and was a Staging Point for units deploying to "Pacific Ocean Theater of WWII" and Korean War. It is located in Pittsburg, CA. It was decommissioned in 1954*

CHAPTER FIVE

LEAVING FT. CAMPBELL KY. TO CAMP STONEMAN CA.

Letter 89 to Letter 101

I am now back across the 38th Parallel Line   Julie Mullenax Van Meter

Letter #89 [8]

December 20, 1950

To: Edith Zinkhan
Frank, WV

From: Sgt John W. Mullenax
On the Train Trip to Camp Stoneman, CA

Dearest "Edy"
    Well, try and write a few lines on this train, but it is hard to do, for it shakes so much. Hon, how are you by this time? I hope these few lines find you OK.
    I have been on this train for three nights now. I have seen some of the most beautiful scenery. I came by St. Louis to Kansas City and hoped to mail this in Salt Lake City. I am supposed to be in Camp Stoneman, California, Friday at 6 o'clock. I had terrific headaches on this trip.
    Darling, I love you and miss you so much. I don't know that I will ever get over being without you.
    Darling, I wouldn't mind making this trip if you were with me on a pleasure trip.
    Well, Darling, I will close and love you more than ever.

Good night and Sweet Dreams, Darling
Love for always,
Love,
Johnnie (Jackass)

---

[8] *Letter was sent in Helper, Utah where the train had stopped.*

Letter #90

December 23, 1950

To: Edith Zinkhan
Frank, WV

From: Sgt John W. Mullenax
Co. B Repl. BN
Camp Stoneman, CA

Dearest, "Edy"
    How are you by this time? I presume you think I am never going to write. Darling, I got in this camp yesterday, Dec 22, 1950.
    It was a long and tiresome trip down here. I saw a lot of beautiful scenery coming down here. I crossed over the Great Salt Lakes. It took about an hour to cross them. Where I crossed them at. It is very nice down here. Everything is so green, although it hasn't been warm.
    Hon, they told me I would be pulled out and put in a Holding Company. On account of my drawing a pension, they said I couldn't draw two government checks. I don't know what they are going to do. They are pulling out all of the ones that are drawing a pension.
    My group of men that I was with are shipping to JAPAN Tuesday or Wednesday. I guess I won't get to go with them, and I sure don't want to.
    Darling, it doesn't seem possible that we have to be apart at Christmas. I hope that this is the first and last. I guess you have a lot of good things prepared for Christmas. I know you wouldn't let Christmas go by without having a lot of good things to eat.
    How is Ben making out? I guess he is happy that Christmas is near. Is Santa going to stop and see him?
    Our mail hasn't caught up with us yet. I hope it does soon. I miss hearing from you.
    Well, Hon, I wish you a very Merry Christmas and a Happy New Year. I love you, Darling, and I hope it isn't too long before we can see each other again.
    Good night my darling,
    Sweet dreams with all my love for always,
    Love, Johnnie.

Letter #91

December 25, 1950 (Christmas Day)

To: Edith Zinkhan
Frank, WV

From: Sgt John W. Mullenax
Co. B Replacement BN.
Camp Stoneman, CA

My Dearest Darling,

    How are you by this time? I just came back from a movie. I didn't know what to do with myself today. I walked all over camp just to put in the day. I just can't realize this is Christmas. One reason you aren't by my side. I miss you so much. You are on my mind at all times. I did have a good dinner today. I ate in the consulate's mess hall, where they feed 1000 men. We had turkey, but it wasn't anything like the meals you prepared.

    I was down in San Francisco last night. I didn't stay long. I am so damn disgusted with everything and myself and so dissatisfied. I am still with the men. I don't know what they are going to do about my pension or how long I will be here. I hope to know in a couple days what they are going to do.

    Darling, I hope you had a swell Christmas. Did Betty and Howard come up for Christmas, or did you have any company for Christmas dinner? How I would like to have been there to help eat the good things, you had cooked. Did Ben like the presents he got and did you get the present I sent you? You may not like that style. I didn't have too many to pick from.

    I washed a lot of clothes today. They have Bendix washers. It costs a quarter to use one and a quarter to use the dryer. It took about an hour and a half to do them. Hon, I love you, and I will never forget you if you change your mind about me.

    Well, hon, I will have to close for tonight, and don't worry about me; I am coming back to you.

    Good night darling, and sweet dreams.
    Love, for always,
    Love Johnnie.

Letter #92

December 27, 1950

To: Edith Zinkhan
Frank, WV

From: Sgt John W. Mullenax
Co. B Replacement BN
Camp Stoneman, CA

Dearest "Edy"
    How are you by this time? Darling, I am getting along OK. By the way, I met Ralph Stone here. He came in a day after me. He shipped from Fort Campbell, KY, to here. I just met him yesterday at a Formation.
    Honey, I don't know what they are going to pull on me about my pension. I have to go to Post Headquarters tomorrow morning. I understand they want me to sign a waiver to give up my pension. I am not going to sign unless I really have to. I will sign a statement to give up my pension as long as I am drawing Army pay. Because I don't think I can draw 2 government checks. I will let you know as soon as I find out what the story is.
    It has been miserable weather down here, raining. I mean a mist about all the time. I have your picture sitting on the stand that is solid. I have been afraid of breaking it. Although I have been careful, but some joker might throw something on it.
    I often gaze at your picture, the most beautiful little woman I know and the one I love.
    Hon, it doesn't look as though this war will end, although I hope it does soon.
    Hon, I will never forget this Christmas and New Year. I did spend 2 in the Army before, but I had no one at home like you, no one to go home to, which makes it a lot of difference.
    Well, Darling, I hope I am not away from you too long. How is Ben getting along? Tell him I said to be a good boy, and I still think of him.
    Good night darling Sweet Dreams
Love for always,
Love,
Johnnie
Love you, Darling.

I am now back across the 38th Parallel Line   Julie Mullenax Van Meter

Letter #93

December 29, 1950 [9]

To: Edith Zinkhan
Frank, WV

From: John W. Mullenax
APO 613
Casual Personnel Dept.
San Francisco, CA

My Dearest "Edy"
    How is my darling by this time? I hope these few lines find you OK. I am getting along OK. I am not doing a whole lot here, only trying to dodge details. I have finished processing for overseas shipment. There is a possibility of my flying over, but I would rather go by ship if and when I have to go.
    Hon, I was down at Post Headquarters yesterday about my pension. I had to sign a paper to the VA that they would stop my payments while I was in the Army. As I understand, when I get out, I will get it back, I hope.
    Darling, I haven't heard from you for 2 weeks. It seems as though it were months. That is the most important thing I have to look forward to here. Hon, I found you a Christmas card, I hope you received it.
    Darling, I am going to send the electric razor home and your picture. I wouldn't want it broken because I wouldn't take anything for it, and where I'm going, it would be more than likely to be broken.
    Well, Hon, I just got back from chow. Ralph and I ate together, and we had a long talk. He told me all the news. He said Doc and Mancehello had a fight when he was back there. Hon, there are a lot of things I would like to tell you, but I don't know how to put them in words. What I mean is I love you more than anything in the world and hope my love for you

---

[9] *Getting ready Camp Stoneman to fly to Camp Drake near Tokyo, Japan*

isn't in vain. I do hope to hear from you soon to know you are alright. Did they get the deer head yet and bring it to you? I guess your friends are still pestering you.

    We are having a Formation tonight. I think it is orders to ship out. I will finish this and let you know after the Formation. Did the people have a big Christmas back there? I happen to think of the night the Waughs took us to Elkins, and I got the hiccups. Hon, I often think of the swell times we had together, and we are going to have them over.

    Well, Honey, I got my orders; I ship out in the morning. I am flying over to JAPAN. I will go to CAMP DRAKE.

    I will say good night for now and sweet dreams, Darling, and take good care of yourself. I will be thinking of you, for I love you.

Love for Always Love,
Johnnie, or Future Hubby.

I am now back across the 38th Parallel Line   Julie Mullenax Van Meter

Letter #94 [10]

December 30, 1950

To: Edith Zinkhan
Frank, WV

From: Sgt John W. Mullenax
C. P. S. – A. P. 613
% P.M. San Francisco, CA

Dearest Darling,
    Will write to let you know my address overseas. So, use this, and I will get my mail quicker. I am rushed this morning, so I won't be able to write much.
    I sent a package yesterday by express to Durbin, WV. To you by COD. Pick it up for me, Hon; I will appreciate it very much; it is some Army clothes, the electric razor you gave me. I won't be near electricity too often, so I decided to send it home. I also sent your large picture. I was afraid it would get destroyed, and I wouldn't want that to happen to a beautiful woman like you, the one I love. I still have you in my billfold. I will say so long until I get where I can write.

    I love you, Darling,
    Love for always,
    Johnnie

---

[10] *This is a farewell letter*

I am now back across the 38th Parallel Line   Julie Mullenax Van Meter

Letter #95 [11]

December 30, 1950

To: Edith Zinkhan
Frank, WV

From: Sgt John W. Mullenax
CPS AP613, %PM
San Francisco, CA

My Dearest "Edy"
    How is my Darling by the time? I hope these few lines find you and Bennie OK. Hon, I haven't heard from you for about 2½ weeks. It sure gives me the blues when I don't hear from you. I know it isn't your fault.
    Hon, I am down in another Company waiting for orders that our plane is ready to fly. I don't know how long I will have to wait. They weigh us and our equipment.
    Hon, if I would have known I was going to be here this long, I would have sent for you. We wouldn't have had long together, but it sure would have meant a lot to me.
    I was in charge of a group today, and a guy said his wife was down at the Service's Club. He wanted to know if he could go to give her goodbye. I couldn't say yes, and I didn't have the heart to say no. I told him to fall at the rear of the troop. When we marched to chow, and not to let me see him. Everything came out OK.
    Well, hon, it will take me about 40 hours to fly to JAPAN. I make 2 stops, one in Honolulu, Hawaii, and one at Wake Island. We are restricted to the barracks.
    Goodnight, Darling. I love you so much. That's words won't tell.

Goodnight,
Love for always,
Johnnie

---

[11] *This is about The Stops he makes to get to JAPAN*

I am now back across the 38th Parallel Line   Julie Mullenax Van Meter

Letter #96 [12]

January 1, 1951

To: Edith Zinkhan
Frank, WV

From: Sgt John W. Mullenax
CPS APO 613 % PM
San Francisco, CA
(En route to Japan, Through Wake Island, where this is mailed from)
(This has Pan American World Airways – en route by "Clipper" – Stationery)

Dearest "Edy"
    Will write a few lines this morning to let you know that I am making the trip OK. It is hard to write. The plane is shaking. I will be in Japan this morning sometime. The captain said we had 8 hundred miles to go yet. We stopped at Wake Island and ate. There wasn't much there. I didn't have much time there. I haven't minded the trip so far. I was up with the pilot. He sure has a lot of things to operate.
    How is everyone back there? How is Ben making out with his dog? I guess you have a time with the two. Tell Ben I said I will take him a ride on a big airplane when I get back. Darling, excuse the writing.
    Hon, I love you and will never change my mind.
    I will close for this time, hoping to hear from you.

With all My Love for Always
Love,
Johnnie

---

[12] *Arriving in Japan, stopping at Wake Island first.*
The US used the airfield and facilities at Wake Island as a key mid-Pacific refueling stop for its mission of transporting men and supplies to the Korean Front.

I am now back across the 38th Parallel Line   Julie Mullenax Van Meter

Letter #97 [13]

January 2, 1951

To: Edith Zinkhan
Frank, WV, USA

From: Sgt John W. Mullenax

Enroute to Japan to Korea
This post card is from, Honolulu, HI

"Hi" Edy
    I landed in Honolulu this morning. I will be here for about 3 hours. It is very beautiful here. The sun rise this morning.
    Darling, hope you are well. Will drop you a line at the next stop.

    Love,
    Johnnie.

---

[13] *This is the paragraph that is on the back of the Hawaii postcard.*

I am now back across the 38th Parallel Line — Julie Mullenax Van Meter

Landed in Honolulu, Hawaii, Jan 2, 1951

Letter #98

January 7, 1951

To: Edith Zinkhan
Frank, WV

From: Sgt John W. Mullenax
CPS APO 613, 5PM
San Francisco, CA

Dear "Edy"

Just a few lines to let you know that I arrived here OK. It took us 36 hours to fly over here from San Francisco. I stopped at Honolulu for 3 hours, and I stopped at Wake Island. I sent you a cablegram. I was limited to so many words.

Hon, I haven't heard from you since I left Kentucky. I sure would like to hear from you to know that you are well and getting along OK.

I miss you so much, darling, and I have to be over here in this darn country. I don't like it here, hon; I would like to be back there with you. How is Ben getting along? Tell him I said hello.

Darling, I guess you are still having your neighbors up there. Have you taken your driver's test yet? I want you to get a driver's license.

Ralph Stone was in California when I left; he didn't know when he was going to ship out. Are they drafting a lot of the boys yet? Hon, I hope this will cease soon so I can get back there. I sure miss those good times with you. Hon, I will never sign anything connected with the Army.

Darling, there isn't much to write about. I have to stay in camp waiting for further orders. How is the weather over there? I guess it is too cold for them to do much sawing. I hope they make some money with the business for you and me. Hon, I am going to go into business when I get out by myself. Of course, I want you as a partner.

Hon, I hope you got that box of clothes I sent; I will pay for it when I get paid. I have been only getting partial pay. They said I would get paid when I get signed to a unit. Then I will send some money.

Love for Always
Future hubby, Johnnie

I am now back across the 38th Parallel Line   Julie Mullenax Van Meter

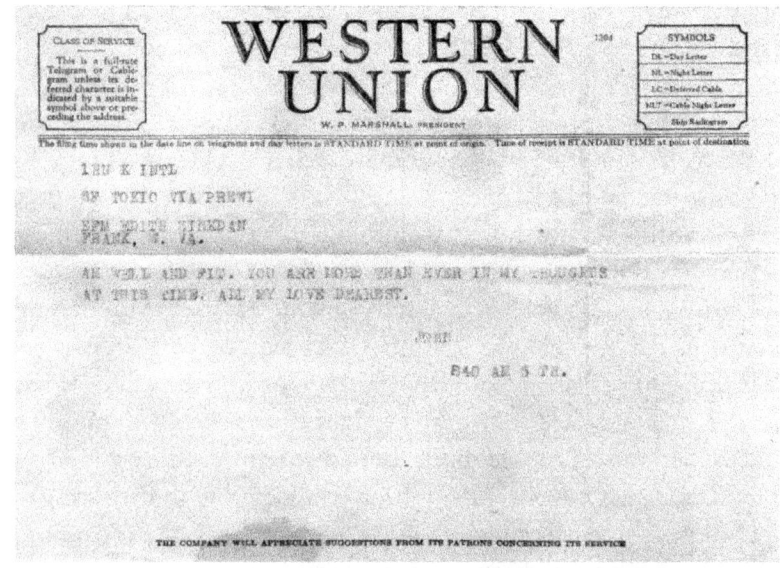

Arrived in Wake Island Jan 6 1951

Letter #99 [14]

Japan
January 9, 1951

To: Edith Zinkhan
Frank, WV

From: Sgt John W. Mullenax
8069 Repl. BN. APO 301
%PM San Francisco, CA

Dearest Darling,
    How are you by this time? Hon, I haven't been able to write every day, but don't feel harsh toward me, for I don't get a chance to write. I still want to hear from you, hon. I am very interested in how you are getting along. You can send me a box if it isn't too much trouble.
    I haven't heard from you since I was at Fort Campbell. How has Bennie been getting along in school? Tell him I want to know if he is still my boy. Does he still have the little dog?
    I sure have been doing a lot of traveling lately. In fact, too much to suit me. I rode the train for a day and a half. I saw Japan from the train. It looked so funny. The rice fields up against the mountains. I can't understand a word they say. I sure don't like it here and never will. Darling, I love you and hoping that it isn't too long before I can be back there with you. Hon, the longer I'm in the Army shows how much you mean to me.
    I am signed to the 25th Infantry Division, but use the address on the envelope until you hear from me. Well, Darling, I am going to have to clean up tonight. I haven't shaved for 3 days. Let me hear from you all the gossip that is going around back there.
    Good night Darling; I love you, and Sweet Dreams.
    Love for Always
    Love, Johnnie.

    P.S.: *I have had dreams. "Haha"*

---

[14] *Assigned to the 25th Division*

Letter #100

Japan
January 11, 1951

To: Edith Zinkhan
Frank, WV

From: Sgt John W. Mullenax
8069 Repl BN APO 301
% PM San Francisco, CA

Dearest Darling,

How are you by this time? Fine, I hope? I am getting along OK. I haven't been doing anything but lying around here waiting for my orders. I got them this morning to be ready to move out. I get on a ship here for Pursan. I will be there tomorrow. I don't know how long I will be there. I will ship out from there (Pursan) to my Company. I don't know where they are, but it will be a Company of the 25th Division. So, I want you to keep up with the news of the 25th Division.

I am in the Infantry, as far as I know. I will be a squad leader of 9 men. This class to the front, we don't get much news. I read a paper last night that was 2 days old, published here in Japan. Hon, I guess it will be some time before I hear from you. I am moving so much that the mail will take a long time to catch up with me. I do know I will appreciate them when I do get them.

Darling, you can send me a box when you have time as if it isn't too much trouble. I would appreciate it very much. Darling, I may expect too much from you. I want you to tell me if I do. But I love you. That is the reason I guess I ask you to do so much for me.

But, when I get back, I will repay you. I never cared for anyone as much as I do you, Darling. I have a lot of time to think about this Army, and it is about you and the wonderful times we spent together. Tell my boy Hello and be a good boy and listen to you.

Hon, how is the stove working? I sure would like to have one of your home-cooked meals now. You are the best cook I know of. I would rather sit at your table than anyone I know of. Hon, did that company fix your breakfast set for you yet? I like that set. You have a good eye for picking things. I enjoyed going shopping with you, in fact, you are the only one I ever went shopping with.

How are Elsie and Fred getting along? Did Roy Lambert get back for Christmas from Alaska? How are Floyd and Ruth getting along? I guess they are still running the gas station. Well, Darling, it isn't long until I move out. I will write you when I get to Korea. I want the future Mrs. Mullenax to take care of herself, for I love her.

Write often, Darling. Love, my Dearest. Love for always.
Love,
Johnnie.

I am now back across the 38th Parallel Line   Julie Mullenax Van Meter

Letter #101

Korea
January 12, 1951

To: Edith Zinkhan
Frank, WV

From: Sgt. John W. Mullenax
Korea

Dearest "Edy"
    Hope these few lines find you well and getting along OK. Hon, I am now in a Korea house. I hope I can get a good night's rest. We have a homemade lamp, a bottle with gas, and a cloth for a wick. It's very poor but will do when I can't get any better.
    We secured a big hill, and some other men took over, and we moved back to our old place. I still haven't heard from you yet. Hope to hear soon. We had a good meal today. I went to a village in front of the lines and got a cow and brought it back for our cooks. I have eaten so damn much rice I might turn into a chink. Ha.
    Darling, I hope this is over soon so I can get back. This is no place here for no man. I am Assistant Platoon Sergeant. I don't know for how long.
    How is Bennie getting along? Tell him I said Hello. How are Floyd and Ruth getting along? Have you been up to Aunt Gertie's lately? Excuse me a second; my electric light is going out. Hon, I don't hear any news here. I don't know what is going on, only a war. I am with a good bunch of men. I am in the 5th R.C.T. attached to the 24th Division.
    I haven't shaved for a week. I hope to get time in the morning. The Medic is here beside me writing to. He is attached to our Platoon. Excuse the writing, darling. Hope you can read this. You haven't by any chance seen my car passing by. They better hadn't been using it. We have a car, and we don't want it beat up. Do you know whether Ralph Stone came overseas or

not. I left him in California. Our interpreter or Korean soldier, just asked me what time he has to guard.

    Well, Darling, I will close for this time and try and get some sleep. Darling, I love you and hope you haven't changed your mind.

    Hope to hear from you soon.
    Love for Always and Sweet dreams.
    Love,
    Johnnie.

CHAPTER SIX

FROM CAMP STONEMAN CA. TO THE PURSAN PERIMETER, S. KOREA

Letter 102 to Letter 169

Letter #102

Korea
Jan 19, 1951

To: Edith Zinkhan
Frank, WV

From: Sgt John W. Mullenax
5th Infantry Regiment APO 24
%PM San Francisco, CA

Dearest Darling,
 How are you by this time? Hope these few lines find you well and getting along OK.
 How is Bennie getting along? I bet he has a time with his dog. Has he taught him any tricks?
 Well, Hon, there isn't a whole lot to tell you about this country. The definition I can think of. It is a stinker hole. I have been sleeping on the ground for a week. I have my sleeping bag to crawl in at nights. It is very hard to keep warm. We only have 2 meals a day.
 Are you still helping Aunt Gertie? Did they get the deer head to you? I left the money at home to pay for it. Well, Hon, I am interrupted; they are calling for me. Darling, I love you and hope that we can share our love together soon.
 Will close for now with all my love.

Love for always,
Love,
Johnnie

P.S.: *Sweet Dreams*

Letter #103

Korea
January 26, 1951

To: Edith Zinkhan
Frank, WV

From: Sgt John W. Mullenax
Co. F 5th RCT, APO 301
%PM San Francisco, CA

Dearest "Edy"
    How are you by this time? I hope these few lines find you well. I haven't written to you for some time; I don't want you to feel harsh toward me. I haven't had time to do anything much. I know that you are anxious to hear from me as I am the same as you. Darling, I haven't heard from you since Dec 16th, 1950.
    By being with my outfit, I should get some mail soon. I just got through washing and cleaning out my helmet. Hon, this is the worst country I have ever been in. In fact, there isn't a thing here but mountain and rice patties.
    How is Bennie getting along? I guess he is growing, so I won't know him when I get back. I have been wandering whether he will pass or not. Tell him I said he had better.
    Darling, excuse the writing, for my fingers are stiff from the cold writing this. We are pushing forward now, which I guess you can see in the paper. We are attached to the 24th Division. I am in the 5th Regimental Combat Team.
    I hope this is over soon so I can get back to you. I don't want no part of the country. Darling, I love you. That is why it makes it hard for me. The other time I was overseas, I didn't care. I never had anyone at home like you.
    Close for this time with all my love.

Love for Always,
Love,
Johnnie, future hubby.

I am now back across the 38th Parallel Line   Julie Mullenax Van Meter

Letter #104

Korea
February 5, 1951

To: Edith Zinkhan
Frank, WV

From: Sgt. John W. Mullenax
Co F 5th Infantry Regiment, APO 301
%PM San Francisco, CA

Dearest "Edy"

How are you by this time? Fine, I hope. How is Bennie? Getting along with his schoolwork. I hope he passes this year. Tell him if he wants that BB Gun, he had better pass.

General Ridgeway was up here yesterday. I didn't get to talk to him.

I have been getting plenty of cigarettes here. If you send a box, don't send any.

I heard a rumor yesterday. We may get relief. I hope it is true. This outfit I am in is from Honolulu.

I just had to stop this, to roll up my sleeping bag to move out in the spare of the moment. That's how it goes here; they keep moving us.

Hon, it is so hard for me to write and not hear from you, although I know you are writing. But my mail hasn't caught up with me as yet. They tell me, here it takes about a month for mail to come and go from the states. I got a haircut yesterday; a South Korean soldier cut it. Looks very good to be cut out in the open.

Well, Hon, I hope to hear from you and to know that I have my Darling waiting for me. I love you, Darling, with all my heart.

Love for always,
Johnnie

I am now back across the 38th Parallel Line   Julie Mullenax Van Meter

Letter #105

Korea
February 7, 1951

To: Edith Zinkhan
Frank, WV

From: Sgt. John W. Mullenax
Co F 5th Infantry Regiment, APO 301
%PM San Francisco, CA

Dearest Darling,

How is my Darling by this time? I am getting along OK.

We have moved back to the reserve of the Regiment. We got back last night and moved the civilians out of a small town and moved in. I just got through shaving and taking a bath. We cut the end out of a 55-gallon gas drum to bathe in. It sure is a treat to get cleaned up again.

I have been living on some of these high mountains. That is where most of our fighting is, taking the high points. I am still hearing rumors. The last rumor I heard is we would move back to Honolulu, where this outfit came from. I heard we would move the middle of next month. I hope. It is time this isn't nothing but a dog's life here.

Hon, I sure would like to hear from you and to know that you and Bennie are getting along OK. I would like to know what is going on back there. I don't hear any news here. Darling, I love you and hope you haven't forgotten me or our promises to each other.

I'm writing outside on a crock where they keep their food. The food sure stinks. I don't see how they eat it. Well. I don't think these people take a bath but once a year, if that. We just had mail call, and one of the new men that came over with me got a package, so I may get some mail from you soon.

I am Assistant Platoon Sgt. I should get another stripe. I hope this means more money.

How is Bennie getting along in school? Has he been listening to you, which he promised me?

My fingers are getting cold. One of the boys is taking a bath in our room. It is small, there are 5 of us sleeping in it on the floor. It sure is hard, but better than sleeping outside. Well, Darling, I will close for this time.

I love you, Darling, and often think of the swell times we have had together. The swellest time any couple could have.

With All My Love, for always
Love,
Johnnie

Letter #106

Korea
February 9, 1951

To: Edith Zinkhan
Frank, WV

From: Sgt John W. Mullenax
Co F 5th Inf. Regt. A.P.O. 301
% PM San Francisco. CA

Dearest "Edy"
 Will write a few lines to let you know that I'm getting along OK. I hope these few lines find you well. Hon, I will be back in reserves for a few days. I don't know just how long, so there isn't anything to worry about while I am back here. Our Company is in a small town, we are staying in Korean houses.
 They heat these houses by building a fire at one end, and the heat goes through under the floor heating the floor. They don't have any stairs. The floor gets so hot sometimes I can't hardly sleep on it. We don't get very good chow here on the front. So, if you could send me something to eat, I would appreciate it very much. Send it by Air Mail.
 Darling, I will send you some money as soon as I get paid. Hon, some of the boys that come over with me are getting mail. I hope to hear from you soon.
 These houses as full of lice. I carry a can of lice powder with me to take care of them. I have to put some in my sleeping bag every night.
 Darling, I hope to get back to civilization soon and back with you where life is worth living. These crummy people, you don't have to worry about me falling for any of these people. I just can't stand them, even the Korean soldiers in our Company. I can't bare them.
 How is Bennie getting along? Does he still have his dog? I bet it has grown up by now. Has he trained it any tricks yet? Tell Ben I said to be

I am now back across the 38th Parallel Line — Julie Mullenax Van Meter

a god boy and I hope to see him before too long. I guess he still likes to go to the Cowboy pictures.

Darling, I love you somehow. I can't put it on paper how I do love you. But actions speak louder than words. Darling, write me often. I know its hard to write when you don't hear from me, but there are times I am not in a position to write.

Tell all "hello" for me, and I do love you,

With all my love
For always
Love,
Johnnie

Letter #107

Korea
February 14, 1951

To: Edith Zinkhan
Frank, WV

From: Sgt John W Mullenax
Co F 5th Inf Reg. APO 301
% PM San Francisco, CA

Letterhead: United States Army Special Services

Dearest Darling,
    Hope these few lines find you well. Hon, I am getting along OK. I am sure getting tired of living out in the field. I am on the Han River looking for Joe Chink. Hope he runs "ha." Is the group still going up to Aunt Gertie's for hot dogs? That is Floyd, Les and Sammie. Tell them I would like to be there too.
    How is Jim getting along? I presume he still lives in the stone house.
    Darling, when I get a letter from you. I will feel a lot better. I know it isn't your fault. I guess someone has gotten my address messed up along the line. Hon, my birthday almost passed without me knowing it. It was just another day, just like Christmas and New Year's. Well, Hon, I am going to run off the paper, so write every chance you get.

    I love you and hope I can prove it soon,

    Love,
    Johnnie

I am now back across the 38th Parallel Line   Julie Mullenax Van Meter

Letters #108

Korea
February 15, 1951

To: Edith Zinkhan
Frank, WV

From: Sgt John W. Mullenax
Co F., 5th Inf. Reg. APO 301
% San Francisco, CA

Dearest Darling "Edy"

    Will write a few lines today. I have got a breathing spell today. In other words, I get a chance to shave and clean up. I have some Korean washing some of my clothes. They were so dirty I believe they would stand up alone. I'm supposed to get a cold shot this afternoon. In fact, the whole outfit is getting them.

    Hon, I got my first letter from you this morning. I have read it over about 5 times. It sure boosted my morale. It was dated the 3rd of January, 1951. The one that you said would be over home for dinner. Now that I have got the first one, they may get to me more regular. Hope so. I'm glad to hear that you and Ben are well.

    Darling, I will try to finish this letter. I was called to chow; the Koreans call it "chop chop" Darling, I haven't been able to write as often as I would like to, so don't feel too harsh toward me. If you don't hear regular sometimes, it is hard to find something to write on.

    I have been doing a lot of walking. I have been going up and down the hills so much I have lost a lot of weight. I think I have been put in for another stripe or for Sergeant First Class. It will be sometime before I know whether I made it. Hope so. For it means more money.

    Hon, you said you were waiting for me, which gives me a greater determination to return when I know I have someone like you back home.

I am now back across the 38th Parallel Line   Julie Mullenax Van Meter

    We got a treat today, 2 Baby Ruth's Bars, at dinner. Hon, if and when you do send a box, don't send anything of value because I can't carry it. I would send you something, but there isn't anything of value here.
    Well, Darling, I love you, and I haven't changed my mind as far as you and I are concerned. Tell all "Hello" and tell Ben I said to be a good boy and hope to see him before lone.

    Will close with all my love for always,

    Love,
    Johnnie

Letter #109

Korea
February 16, 1951

To: Edith Zinkhan
Frank, WV

From: Sgt, John W. Mullenax
Co. F 5th Inf. Regt, APO 301
%PM San Francisco, Ca

My Dearest Darling,

    How are you by this time? I am getting along OK. Hon, I had a great surprise last night. I received a box from you. It was the one with the cards and candy. The candy sure was good. My lieutenant and buddies sure bragged about it. They wanted to know who sent it. I told them about my future wife. They said you sure can make candy. I told them I had known it for quite some time. Oh, the cards. The boys almost broke my arm, grabbing them to see them.

    Hon, I couldn't understand there were 2 wrappings on the package with stamps and address on both wrappings. I thought it could be a wrapping off of another package.

    Darling, I know that it costs you plenty to send these packages, but I will send you some money as soon as I get paid. I hope to get paid next month. It has been snowing and blowing all day.

    From what I heard, I will move back to the front tomorrow. I sure hate sleeping in these foxholes. I just hope this is over soon so I can get back to you. I didn't know how much I would miss you. Darling, this sure lets me know how much we mean to each other by being apart and so faraway. Darling, I sure appreciate the package and hope to pay you someday somehow for what you are doing for me.

    Well, Hon, by looking at all these cards, it's about time for me to have a dream. "Ha," Hon, you said they never brought the deer head. I can't understand. It should be finished by now. Did you receive the watch from Fort Campbell and the box of clothes I sent to you COD from me?

Well, hon, I will have to close. Tell all Hello and tell our little man I want him to see after things around the house by helping you. He never did tell me what he got for Christmas. I love you, Darling, and I know you love me, and I trust you.

>Love for Always,
>Love,
>Johnnie (future hubby)

Letter #110

Korea
February 20, 1951

To: Edith Zinkhan
Frank, WV 26807

From: Sgt John W. Mullenax
Co. F 5th Inf. Regt. APO 301
% PM San Francisco, CA

Dearest "Edy"
    Here comes a few lines to let you know that I am getting along OK. I hope these few lines find you well and getting along OK. Hon, I received nine sweet letters from you yesterday. Some were in December and January. They were nice to receive even if they were late.
    I will give you some idea of where I am at. I moved out of reserve to the front and walked miles and miles over mountains in one day. I only got one meal, and I am now sitting looking over the Han River at the Chinks.
    One of my men got a box with some candy which sure helped out. They dropped us rations yesterday. I have lost a lot of weight.
    Darling, I hope this is over soon, so I can get back to you and Ben. I heard some rumor we may stop at the 38 Parallel Line. I hope so.
    Darling, I love you, and you don't have to worry about me having anything to do with these women. They are dirty, and I sure don't like their looks. Well. My pen ran out of ink, so I had to finish with a pencil. I hope I get to get out with a jeep tonight. I think I can.
    Well, I love you as much as ever, and I don't want you to doubt me. I will close for this time. Hoping to hear from you often. I have been pushing for 4 days and haven't had time to write, hardly time to sleep.

    With all my love.
    For Always,
    Love,
    Johnnie

Letter #111

Korea
February 22, 1951

To: Edith Mullenax
Frank, WV

From: Sgt John W. Mullenax
Co F 5th Inf Regt. APO 301
% PM San Francisco, CA

My Dearest "Edy"
    How are you by this time? Hope these lines find you well. Hon, I received nine sweet letters from you yesterday, which made me happy. I received and appreciated them so much.
    Darling, I have been on the front for 5 days. They pulled us back last night, about 2 miles. I have been taking it easy for today. I have been going over mountains' night and day, and I sure have walked off a lot of weight.
    I also received a good many letters, from Mother and also one from your mother. She said she wanted to stay with you a couple months before Pits started teaching school and that he was going to teach in that county.
    You said that they brought the deer head. Would you take a picture of it and send it to me. I am glad you like it, so wherever you think is best to hang it suits me.
    Hon, I got the box of candy which I think I told you about in one of my other letters, it sure was good. One of the boys got a box which sure helped out because we weren't getting much to eat.
    You said some of them were over for dinner and that Dad liked your cooking. He is like his son; he knows a good cook. Darling, if I just could get some of your cooking now.
    Hon, would you take a picture of yourself and send it to me. I want to see if you are as beautiful as when I left you. Which I know you are, but please send me a photo of yourself. I was getting a lot of Mullenax

I am now back across the 38th Parallel Line   Julie Mullenax Van Meter

Mail. One I got yesterday was for Capt. John T. Mullenax stamped at Connecticut. And 2 for S. A. Mullenax stamped North Carolina. I turned them back in.

    Darling, I love you, and you aren't wasting your time by waiting for me because I am sure coming back. All I hope it isn't too long to wait.

    Will close for now with all my love to the sweetest girl I know.

Love for always,
Love Johnnie, Hubby

Letter #112

Korea
February 1951

To: Benjamin C. Zinkhan III
Frank, WV

From: Sgt John W. Mullenax
Co F 5th Inf Regt. APO 301
% PM San Francisco, CA

Hi Buddie,

    How is my boy by this time? I received your nice letter, sure was glad to know that I still had my pal.

    Are you taking care of mommy like I told you and listening to her? It will please me if you pass this year. How about working on those lessons more.

    Bennie, I hope I'm not over here too long. For you and I can have a bit of fun fishing and swimming. Have you taught the dog any tricks? I'm glad you got a lot of nice things for Christmas.

    Where I was, I couldn't get you anything, so when I get back, I will bring you something.

    Write often,
    Love,
    Johnnie

I am now back across the 38th Parallel Line Julie Mullenax Van Meter

Letter #113

Korea
February 19, 1951

To: Edith Zinkhan
Frank, WV

From: Sgt John W. Mullenax
Co F, 5th Inf Regt. APO 301
%RM San Francisco, CA

Dearest "Edy"
    Will drop you a few lines to let you know that I am getting along OK. Hope these few lines find you well and getting along fine.
    Hon, I am pulled back off the lines for a few days, don't know for how long.
    We are getting only two hot meals a day. They are supposed to be hot but are cold when they get here. The roads are bad, that is the reason.
    Sometimes we get a can of C rations for a third meal. We are in Koreans houses which isn't much of a house but is still better then out in the cold and rain.
    How does the war news sound back there? We heard the Russians are moving in China. Hope they don't come on down here. What beats me, we had a rifle inspection here on the front.
    Oh yes, I was offered 10.00 for those cards you sent me. By the way, where did you get them? I don't like to sell them because they are from you.
    Darling, I have been getting a lot of lovely letters from you, sometimes 7 a day. I am sure happy to receive them and to know that I have my Darling waiting for me.
    Hon, getting your letters and boxes sure helps me out here. It gives me something to look forward to. I get letters from Mother. She doesn't tell me too much about the sawmill. I guess the men don't tell her what they are doing or the things I would like to know in the progress of it.

Darling, keep your chin up, our day should come soon. I think I told you I received a letter from your mother. She doesn't like her job and is talking of quitting. Hon, we had a very good breakfast this morning. Gravy on bread with scrambled eggs and coffee, but cold, it got here about 10:30 o'clock.

Well, Hon, I will have to stop because they are hollering for me. I think we have to take a hike somewhere this morning. They said it would take about 3 hours. Well, Darling, keep those sweet letters coming, for I appreciate them so much and to know that I have my darling to depend on.

Darling, tell Johnnie Bennett to bring the cap down to Floyd's, and you can pick it up there. I had forgotten about it. Tell all Hello and that I am getting along OK. Darling, I love you, and you don't have to worry about me doing something I shouldn't. I haven't seen anything that would be appealing to me.

> Will close with all my love.
> Love for always,
> Love,
> Johnnie

I am now back across the 38th Parallel Line   Julie Mullenax Van Meter

Letter #114

Korea
February 27, 1951

To: Edith Zinkhan
Frank, WV

From: Sgt John W. Mullenax
Co F 5th Inf Regt. APO 301
%RM San Francisco, CA

Dearest "Edy"

    Received a letter from you this morning. Hon, I am so glad to receive those sweet letters from you and to know that you are alright.

    Darling, it seems as though we have to suffer to get what we want or want to accomplish. Some say you and I will have things as we want them. Darling, I will try and finish this before dark because I can't see to write after dark.

    I started this letter this morning and had to go out and work a road into a small valley where we are located. Of course, I didn't do much. We have Koreans to do the work. It has thawed out a lot here and is plenty soft.

    Darling, I still have the pen you gave me, but can't carry it along. Sometimes I get some when I get back to the reading camp. Darling, I am still along the Han River we are holding here. I think they are closing in on the sides.

    Darling, I have received 2 boxes, a box of Hershey's and a box of homemade candy and cards.

    What does Doc think or say about being a father? Hon, don't forget to send me the pictures I wrote about. Send them as soon as you can. Darling, it may be too much me asking you to stay there. For you and I was always going somewhere, I can imagine how lonesome you get. I have been hearing something about a rotation plan for the troops here in Korea. I don't know just how the points works. But hope I can soon get enough.

I am now back across the 38th Parallel Line   Julie Mullenax Van Meter

Darling, they are hollering again. It seems as though every time I start to write you, they start. I will close for this time, hoping this finds you and Bennie well.

Darling, I love you, and you are the only one if we are a long way apart. You can still trust me and depend on me.

I often think of the swell times we have had together and hope we can have these times back.

I love you for always,
Love,
Johnnie, future hubby

I am now back across the 38th Parallel Line   Julie Mullenax Van Meter

Letter #115

Korea
March 2, 1951

To: Edith Zinkhan
Frank, WV

From: Sgt. John W. Mullenax
CO. F 5th Inf. Regt. APO 301
% PM San Francisco, Ca

Dearest Darling,
    Just a few lines to let you know that you are always in my thoughts. I love you, Darling, that is why I am always thinking of you. We had a beautiful sunrise this morning.
    It is seven o'clock. My men are sleeping, so I came behind the hill and built a fire and wrote you a letter. It is cold on the fingers this morning. Darling, I received the Birthday and Valentine cards yesterday evening they sure were beautiful. I was so glad to receive them, and I also got 2 sweet letters from you. Darling, you were asking me whether or not you should go to work. Well, Darling, I rather you wouldn't, but I will leave it up to you. I don't want to be busy or go against your will.
    I would like you to be home when I get back. Hon, I heard a rumor that soldiers that have been here 6 to 8 months get to go home. I have already got 2 months. But that is going to be a long time for me and you to be apart. Don't give up hopes, Hon, I haven't given up yet.
    I know that it gets lonesome there for you, but please, bear with me. There are a lot of things. I am lonesome for over here. I am wearing the same clothes I left Japan with. I don't get to wash very often, and living out in foxholes and around a swamp can get a person dirty.
    Sometimes, I get so damn mad at myself for getting into this, and I know it wasn't no one's fault but my own.
    Tell Bennie the Valentine's, was nice and I liked it a lot.

I am now back across the 38th Parallel Line   Julie Mullenax Van Meter

    Darling, I love you, and what I want most is to get back to you so that we may share our love together. Write me often.
    Hon, and send me a box a week if you can.
    My chow just got here, so will close for this time.

    With all my love, to the sweetest girl I know.
    Love for always,
    Johnnie, Jackass.

Letter #116

Korea
March 4, 1951

To: Edith Zinkhan
Frank, WV

From: Sgt John W. Mullenax
Co F 5th In Regt. APO 301
%PM San Francisco, Ca

Dearest Darling "Edy"
    Will drop you a few lines to let you know that I am always thinking of you. Darling, I hope these few lines find you and Ben well and getting along OK. I haven't received any mail for 3 days now, but that don't bother me too much because I had to wait so long for the first.
    I still like to hear every day. Hon, I got partial pay, so enclosed you will find a money order for $60.00 less the fee. Darling, buy Ben an air rifle because I didn't get to get him anything for Christmas. The balance, you can buy yourself a dress or whatever you see fit to use it for because it is yours. For Darling, you have done so much for me, sending me boxes and Dad and Mom's Christmas presents. In other words, I want you to have the money for your own use. Tell Ben that it is his Christmas present from me.
    Darling, I am in the front again. I have been here for 3 days and have been having a hard time so far. I talked with the Sgt. And he said I was at the top of the list for Sergeant First Class. Hope I make it because it means more money. I haven't had much rest for the last few days until today. I got a few hours' rest and time to drop you a few lines.
    Hon, there is a rumor that if a man is in Combat 6 months or in Korea, he will get rotation. Hope it is. So far, I have 4 more months. Hope it is true. Darling, I am running out of news, so I will have to close for this time.
    Darling, I love you. I can't tell you too often, because it is true. I hope to get back so we can share ours together.

I am now back across the 38th Parallel Line   Julie Mullenax Van Meter

    Darling, write often, for that is all I have to look forward to over here.

    With all my love for always,
Love, Johnnie

I am now back across the 38th Parallel Line	Julie Mullenax Van Meter

Letter #117

Korea
March 10, 1951

To: Edith Zinkhan
Frank, WV

From: Sgt. John W. Mullenax
Co F 5th Inf Regt. APO 301
% PM San Francisco, CA

Dear Darling,

 Will drop you a few lines to let you know that I am getting along OK. Hope these few lines find you well. Hon, I haven't had the time or wasn't where I could write you for quite some time. Hope you aren't sore at me because I know what a few lines mean to you because I know what they mean to me.

 I hadn't heard from you for about a week, it was because of the road, we couldn't get hot chow all the time. I am back off the lines for a few days, and I got some sweet letters from you and 2 boxes, a box of cookies and a box of Hershey's. Boy, the boys went crazy over the cookies. They said to tell you how good they were.

 We moved back to the village and still had to sleep outside because there is smallpox in the village. I have just got shaved and washed, which I sure needed.

 Who is moving in the houses of Floyd's? Why are the rest moving out? Darling, I am sure glad to get the boxes. Mom said she was sending me a box or that it was on the way.

 It feels good today; It has warmed up, and the sun is shining. Hon, I hope this is over before too long. This is a hell of a hole here, and I long to be back there with you, Darling. It looks like the ER are getting a raw deal all around. When I get out of this, they sure won't get me again.

 Darling, I sent you a money order, let me know when you get it so I will know it didn't get lost.

We are pushing the chinks. I am across the Han River. I tried to cross the Han once, and they pushed us back, so now I am across, and 2 of our Battalions are pushing on.

Darling, I hope the boy passes because one year means a lot. Tell him I said Hello. Darling, I love you and hope I am not away from you for too long. So, I will close for this time.

Loving you always,
Love and sweet dreams, my darling,
Johnnie

Korea, 1951
Sgt. John W. Mullenax standing on the right

I am now back across the 38th Parallel Line   Julie Mullenax Van Meter

Letter #118

Korea
March 11, 1951

To: Edith Zinkhan
Frank, WV

From: Sgt. John W. Mullenax
Co F 5th Inf Rgt APO 301
%PM San Francisco, CA

Dearest "Edy"

How are you by this time? Fine, I hope. I am getting along OK. I am sleeping in a pup tent once again along a river against the hill. It is a cloudy and dreary day today. Hope the sun shines long enough to dry the clothes I have washed.

Hon, I got another box from you, the one with the nuts and candy in it. I have received 5 boxes from you by now. "Thanks a lot". I sure appreciate them, and my men do also because I divide with them. They divide with me when they get one. I wrote you a letter yesterday and one to Mother and Geneva. I am catching up on my writing. I also want to write one to your mother today. I haven't answered her letter yet.

Darling, I hope we get to stop at the 38th Parallel. They haven't said as yet, whether we will cross or not. We have the chinks on the run. Hope they keep running. I had my picture taken with my men when we crossed the Han River. I was on the first boat and got pushed back. I mean, a newsman took it. Don't know whether it will come out in the paper or not.

Darling, I love you, I think you should know because I have tried to show you for some time, and I believe you, darling; I just hope I am not away from you too long. Write often, darling, that is the only thing I look forward to every day. I will close for this time with all my love for always.

Sweet Dreams,
Love always,
Johnnie.

P.S.: I am not about to change my mind.

Letter #119

Korea
March 12, 1951

Dearest "Edy"

    Will drop you a few lines to let you know that I am getting along OK. I hope these few lines find you well and getting along OK. I am still on the riverbank in reserve of the Regiment. Don't know how long we will be here, but when I leave here, I will be climbing mountains again, and they are sure steep.

    Hon, I will try and finish this letter today. I received 2 letters from you which are important for me to get far. I am so glad to know that I still have my Darling back there. Darling, you can't write that you love me too often, for that is what I want to hear, for I love you, Darling.

    Darling, they have set up some rotation plans for the troops to the states. I don't know how long you have to be here. I think it is 6 to 8 months.

    I only hope so far that is a lot to look forward to. I want, most of all, to get back to you. Hon, I also got a letter from Floyd, sure was glad to hear from him.

    Hon, I also got to take a shower which I think was great. They set it up along the river, in fact, it was the first shower I had had in a couple months. Hon, I have been getting your sweet letters for 10 to 12 days, and I am so glad to hear that you are well and getting along fine.

    Darling, I am loving you for always, and please don't worry about me changing my mind, for my mind is made up.

    Write often, Darling, for you are my future.

    Love for always,
    Love,
    Johnnie

I am now back across the 38th Parallel Line

Letter #120

Korea
March 15, 1951

To: Edith Zinkhan
Frank, WV

From: Sgt. John W. Mullenax
Co F 5th Inf Regt. APO 301
% RM San Francisco, CA

My Dearest "Edy"

    Here comes a few lines hoping that they find you well and getting along OK. Hon, I am still in Reserve. By reserve, I don't mean we don't do nothing; we have something to keep us busy getting our equipment repaired. Of course, we get sleep and rest, which we wouldn't get up on the line.

    For instance, I was Orientated by the captain today on the map of our objective. Which we have to take; I have to be able to find myself on the map where I am on the ground, so when I have to let them know just where I am at. We have a code when we call over a radio given this information. I understand we will move out tomorrow or the next day; of course, that isn't definite, and it could be changed at any time.

    Darling, I haven't heard from you for 2 days. Hope to get a letter tonight. Hon, I have lost so much weight over here. Darling, the rotation is in effect. I understand that they will start the rotation for men on April 1 from this Company.

    That gives me something to look forward to. I may get back to you in 6 or 8 months. Darling, I love you; words can't express my feelings toward you. Don't worry about me doing anything foolish. I am just counting the days to get out of here and back to you and your love which I cherish so much.

    Hon, I sent you a money order. Let me know when you receive it.
I will be loving you no matter how far or where.
Sweet Dreams, Darling,
Love for always,
Your Lover, Johnnie (Jackass).

Letter #121

Korea
March 17, 1951

To: Edith Zinkhan
Frank, WV

From: Sgt. John W. Mullenax
Co F 5th Inf. Regt. APO 301
% RM San Francisco, CA

Dearest Darling,

Here comes a few lines to let you know that I am still OK and thinking of you more each day and loving you. Darling, I hope these few lines find you and Ben well and getting along OK. I am so anxious to get the pictures you took. I received 2 back letters last night from you although they were late. I sure was glad to get them.

Darling, I just got word we move in the attack tonight or in the morning. Hon, I only hope this don't last so long. I never was in a place that was so near to nothing as this. These damn people never wash and they eat a mixture of food that I can smell for a mile and we have Korean soldiers right in our company. So, you see what I mean.

We haven't had definite order to cross the 38 parallel line. I hope not.

Darling, the rotation first of April. You have to have six to eight months in Korea. I have 2 months now. Darling, I will close for this time. I have to get ready to move. I love you, Hon, and hope we aren't apart too long.

Love for always,
Love Johnnie, (Jackass)

Letter #123

Korea
March 16, 1951

To: Edith Zinkhan
Frank, WV

From: Sgt John W. Mullenax
Co F 5th Inf Regt APO 301
% PM San Francisco, CA

Dearest "Edy"

Received a letter and an Easter card from you. I am so glad to know that you are well. Darling, the Easter Card was really nice.

Hon, I have no way of purchasing any cards here. You ask me about writing paper. Well, Hon, I am getting all I need in an area like this, but when I go up online, sometimes I can't get any nor, I can't carry any with me. It would be destroyed, for I have no suitable place to carry it. Sometimes we get those folding paper envelopes to combine sent up by the company. I expect to go online in the near future.

I will also write whenever I get a chance. There are a lot of times we stay on a mountain for a week.

Darling, I also received a box last night, the one which had cookies and donuts in it. The boys said you were a swell cook, for they sure went for those cookies. You are a swell cook and a loveable woman.

Darling, I love you; I have for a long time and sure have spent the swellest times of my life with you. Darling, I will close for the time with all my love,

Sweet dreams, darling.

For always love,
Johnnie.

Letter #122

Korea
March 26, 1951

To: Edith Zinkhan
Frank, WV

From: Sgt John W. Mullenax
Co. F 5th Int Regt. APO 301
% RM San Francisco, CA

Dearest "Edy"

Hope to get this wrote before it gets dark, for I haven't written you for 8 days. I haven't had a chance to write. Our Company had a hard time, we were in the mountains. How are you and Bennie getting along? Fine, I hope. Hon, I have been getting your sweet letters and boxes. I got a box with homemade candy and it was sure good. The men said you were a good cook.

I also got the pictures of you and Ben and you look as lovely as ever. Hon, I just got back from the Red Cross; they wanted to know if I had been getting my mail yet. I explained that I had. Darling, I am glad to know that you are that much interested in me.

Darling, I think I will be in reserve for a few days. So, I may be able to catch up on my writing. I will have to close for this time. Darling, you don't know how I would like to hold you in my arms again.

I love you, Darling. Write often, for that is my hopes for another day.

Love for always,
Love,
Johnnie

P.S.: *Why should Mona Bell call the future baby Johnnie. I never had nothing to do with that.*

I am now back across the 38th Parallel Line Julie Mullenax Van Meter

Letter #124

Korea
March 27, 1951

To: Edith Zinkhan
Frank, WV

From: John W. Mullenax
Co F 5th Inf Regt APO 301
% PM San Francisco, CA

Dearest Darling "Edy"

  Hope these few lines find you well and getting along OK. Hon, I am not sure how many boxes, but I think I have received seven boxes. I sure appreciate them very much and so do the men in my Squad.

  Hon, you look like the same little Darling I left and as sweet as ever from the picture you sent. Tell Ben I was glad to get his picture. It looks like he has grown a little man and he has a nice dog. Hon, I never got the pictures of me and you that you said you sent. It has started to rain again. I got wet the night before last night and it is miserable.

  I told Dad to pay the note off that I had with the $50.00 I was sending home a month. I sure would like to have a new Buick. Hon, that is if we can afford, and I am saving every cent over here, only a little for PX rations we get once in a while. Hon, the candy is sure good. Darling, I don't know what I would do if I didn't have you, for you are the one that is keeping me going over here.

  Darling, how I would love to hold you in my arms again, for I am so lonesome for you and your companionship.

  Darling, I heard a rumor that the ER would only have to pull a year. Hope so, but we hear so much over here of rumors. I hope we stop at the 38 Parallel line, for I am not too far from it. Hon, do you think that you can handle the garden. I mean, can you get it plowed and etc.

  Well, Hon, I have some work to do, so I will close for this time.

I am now back across the 38th Parallel Line   Julie Mullenax Van Meter

    Wishing you Sweet Dreams, I love you, Hon, and there is no doubt in my mind.

    Love for always,
    Love,
    Johnnie.

Letter #125

Korea
April 1, 1951

To: Edith Zinkhan
Frank, WV

From: Sgt. John W. Mullenax
Co. F 5th Inf. Regt. APO 301
%PM San Francisco, CA

Dearest Darling,

 Here comes a few lines to let you know that I am OK and still thinking of you and love you more than anything in the world. I was never going to write. I haven't been able to write since I wrote those last 2 letters when I was off line for 2 days. I have had hard days since then.
 I am on a mountain holding here on the 38 Parallel line.
 Honey, I sure hope we stop here, for I have had enough of this. I have been living in the mountains about all the times.
 I can say it is a lot warmer today, about the nicest day we have had since I have been over here. Hon, I haven't heard from you since those 2 days I was off line. I may have mail at the rear CP (Command Post).
 I hope so that it keeps my morale up. Darling, I have been getting your boxes. I am sure glad to receive them and so are my buddies. Darling, I am saving all my money and I told Dad to pay off that note at the bank I had. So, you and I will be able to start off not owing anyone.
 I sent you a money order, let me know if you got it or not. I will know to report it. Darling, I hope these few lines find you and Ben well.
 I love you with all my heart.

 Love for always,
 Sweet dreams,
 Love Johnnie, Hubby.

I am now back across the 38th Parallel Line   Julie Mullenax Van Meter

Letter #126

Korea
April 2, 1951

To: Edith Zinkhan
Frank, WV

From: Sgt John W. Mullenax
Co f 5th Inf Regt apo 301
% PN san Francisco, CA

Dearest Darling,
 Hope these few lines find you well. Hon, I received 3 letters from you today and how I am glad to get them. I know how you feel when you don't hear from me. Don't feel harsh towards me.
 Darling, I am sure I will write when I possibly can. I am now on the 38 Parallel line. This makes the second day here. I wrote you yesterday. I hope and pray that we don't have to go any further than this. For I am sick and tired of these mountains. I sleep and eat on top of them.
 Darling, I have been getting your sweet letters and boxes. Hon, you can't have any idea how I long to hear from you sometime. It is some time before I get your mail, but the reason is they can't get it out to us then. I get it all at one time. By the way, I got a package from Joe today. It was a small package of gum and different kind of candy bars. Darling, I love you and long for you more each day.
 We have lost a lot of time together, which I am sure we can make it up for. Darling, tell me all the news, for I don't have any way of hearing or reading any here. I will close with all my love for always and hope we can share our love together soon.
 Sweet dreams.
 Love for always,
 Love,
 Johnnie

 P.S.: *I just got a box that came with fudge milky ways and peanuts. Thanks, Hon.*

I am now back across the 38th Parallel Line   Julie Mullenax Van Meter

Letter #127

April 6, 1951

To: Edith Zinkhan
Frank, WV

From: Sgt John W. Mullenax
Co F 5th Inf Regt. APO 301
% PM San Francisco, CA

Dearest "Edy"
    Darling, I will try and write a few lines today, for I haven't written you for a few days. I am on the front line on a big high mountain. I have crossed the 38th Parallel line. I was in hope we wouldn't cross, but I guess we will have to go on. The fighting is getting harder. I heard a rumor that we would go in reserve tomorrow.
    I hope so, for I have a long beard and haven't been able to wash for some time. I haven't gotten any letters from you for a few days. I guess they are back at the rear CP. I will get them when I get off this mountain.
    Mother wrote and said I had 2 new nieces. Fred & Geneva (Sister). I sure was surprised at Geneva. I guess I am an Uncle again, but still ain't a pop. Ha.
    Darling, I hope these few lines find you and Ben well and getting along OK. How about the garden. Are you going to tackle it or not? I realize that it is too much of a job for you alone. Hon, I have been getting your boxes and You can't realize how much I appreciate them.
    Hon, are you and your friends getting along? What is Ralph Stone's address? I may drop him a line sometime. Is Ben going to pass this year? I hope so. Did you get him the gun? Tell him to be careful with it. Darling, I sure like to get your sweet letters, for I love you and thinking of you always and I often think of the wonderful times together and hope we can continue soon. I will close with all my love for always.
    Sweet dreams, Darling,
    I love you.
    Love for always,
    Love, Johnnie

Letter #128

Korea
April 12, 1951

To: Edith Zinkhan
Frank, WV

From: Sgt John W. Mullenax
Co F 5th Inf Regt. APO 301
% PM San Francisco, CA

Dearest Darling "Edy"
    How do these few lines find my Darling by this time? Fine, I hope. Hon, I am getting along OK.
    What I mean is the conditions I have to put up with. I am now in reserve after being online for 23 days. I think we deserve to be back for a few days. I don't know for how long we will be here. I want to catch up on my letter writing. I got announce of Geneva's baby. I sure was surprised. She said Rosco was as proud as a peacock. I don't blame him.
    Hon, when I got back to the rear CP, I had 3 packages waiting for me. Two were from you and one from my mother. Hon, I sure appreciated them and everything was good; we were hungry. I passed the homemade candy out among my men and they went wild over it. I almost didn't get any myself. I told them it was from my woman and they said you are our woman from now on.
    Darling, we all have something to look forward to now. Rotations have begun. Yesterday and we are supposed to send some more in a few days. Darling, I think I will be due to come home in three months. I hope there isn't any change in the plans. I may be able to get my rating now, for I have been acting Master ever since I have been over here, but they were full up on the ratings and couldn't give me mine, although I was doing the job.
    Darling, I have been getting your letters. Sometimes I get a bunch of them, for they hold them until we get down and out of the hills. Hon, I am so glad to hear from you and to know that my darling is OK. How is

I am now back across the 38th Parallel Line   Julie Mullenax Van Meter

my boy getting along? Tell him I said hello and be a good boy and listen to Mother. Well, Darling, I am going to have to stop for this time, for I hear someone hollering my name. I am hidden to write so they won't bother me. Darling, I love you and hope we can be together before too long.

>With all my love for always
>Always love,
>Johnnie

I am now back across the 38th Parallel Line   Julie Mullenax Van Meter

Letter #129 [15] [16]

Korea
April 14, 1951

To: Edith Zinkhan
Frank, WV

From: Sgt John W Mullenax
Co F 5th Inf Regt. APO 103
%PM San Francisco, CA

Dearest Darling "Edy"

Hon, I received 3 letters from you last night and I also got a box, the one with the peanuts and homemade candy. Darling, I am so glad to hear from you and to know that I have my Darling waiting for me. That candy was so good I almost got sick, I ate so much. So, does my buddies think it was good. They said to tell you it was the best homemade candy that they had eaten.

I was just informed by my captain that I am to take over the Third Platoon.

Hon, I am still in Reserve. I don't know how long I will be back here. But it sure was a pleasure to get down out of the mountains and get a bath and clean up. Once again, I am also getting 3 hot meals here, for we are right beside our kitchen. Hon, I am going to my first movie in about 4 months. We have a movie at Bataan in the field and the First Sgt. wants me to go with him. They said it was a shoot 'em up, so I may be trying to dodge the bullets.

Hon, since Ridgeway has taken over, there may be some change in this.

With all my love for always
Always love,
Johnnie

---

[15] *Informed by Captain to take over the 3rd Platoon. Commander Colonel Ward has inspected the Company.*

[16] *General Matthew Ridgway (3-3-1895 / 7-26-1993). General Ridgway replaced General Walton Walker, who died in a vehicle accident, as Commander of the 8th Army. April 1951 President Truman relieved General MacArthur and replaced him with General Ridgway who oversaw all UN forces. General Ridgway left Korea in May, 1952.*

I am now back across the 38th Parallel Line   Julie Mullenax Van Meter

Letter # 130 [17]

Korea
April 15, 1951

To: Edith Zinkhan
Frank, WV

From: Sgt John W. Mullenax
Co F 5th Inf Regt APO 301
% PM San Francisco, CA

Dearest "Edy"
    Here goes a few lines to let you know that I am OK. I hope these few lines find you well and getting along OK.
    Hon, today was a nice warm day. It sure is different than those cold days we spent. It still gets chilly at night. Darling, these men that are due to go home are a happy bunch. I know just how they feel.
    I am looking forward to that day of getting out of this hole. We are getting a lot of replacements now. I hope they decide to discharge the ERs. I haven't heard from you for a couple of days. A lot of times, we get as many as three letters in one day.
    Hon, I am in my new Platoon, and I am trying to get acquainted with the men. I just got a roster made of the platoon.
    Darling, I love you and you can't realize how I long to be with you. It seems like we have been apart longer than it has been. I hope it isn't as long as it has been. I went to the movie last night, it made me think more of home, for I hadn't seen a movie for so long. Darling, you sure will get to do a lot of cooking when I get home; I have lost so much weight. I still think of the good meals you used to cook.

    Sweet dreams,
    Love for always, Darling,
    Love,
    Johnnie

---

[17] *Getting to know 3rd Platoon, Received new roster*

Letter 131

Korea
April 17, 1951

To: Edith Zinkhan
Frank, WV

From: Sgt John W. Mullenax
Co F 5th Inf. Regt. APO 301
% PM San Francisco, CA

Dearest Darling "Edy"
    Will write a few lines to the sweetest person I know and hope that you are well and getting along OK. Hon, I am still in reserve and how I enjoy being back here off the line. I am sleeping in a pup tent, it gets chilly at nights. Yet, but warms up during the day. It has been windy today and the dust has been terrible.
    Hon, there is a happy bunch of men that has the time in over here and are up for rotation. I am looking forward to that day when I can get back to my darling. I haven't heard from you for four days. I guess I will get a group of letters at one time. Hon, you don't know how happy I am to hear from you. I heard that rumor that they are setting up a three-day rest at Soul, Korea. I mean there will be so many out of a Company at one time, I hope I can get this soon if it's true.
    I was down to get a shower today and how I enjoyed it. I will be glad when I can get back in the bath tub once again. Monna is sending me the Highland Recorder. Now it gives me the news around home. Darling, I love you and long to hold you in my arms and know you are mine. I answered your mother's letter. I bet she thought I was never going to write. But will say this, you come first on my list of letters.
    Darling, I often think of the wonderful times we have had together. I hope in the near future we can continue these enjoyments. Have you planted a garden and are you still keeping the flower garden? How is my

I am now back across the 38th Parallel Line   Julie Mullenax Van Meter

boy behaving? Good, I hope. Darling, I have run out of things to write about. Only I could write about us all day.
    I will close loving you more each day.
    Sweet Dreams
    Darling for I love you,
    Love for Always
    Love,
    Johnnie

I am now back across the 38th Parallel Line   Julie Mullenax Van Meter

Letter #132 [18]

Korea
April 18, 1951

To: Edith Zinkhan
Frank, WV

From: Sgt John W. Mullenax
Co F 5th Inf. Regt. APO 301
% PM San Francisco, CA

Dearest Darling "Edy"
    Hon, I received three sweet letters from you last night. You can't realize how glad I am to hear from you and to know that you are well. I received those 2 pictures of you and I. I can see that my darling is as beautiful and as sweet as ever.
    Hon, I am a Platoon Leader holding the job of the 1st. Lt. and I am only a Sgt. I have been raising hell about it, so they may start to step me up. I have to go up by steps. Hon, all I want is to get out of here, but if I am doing the job, I expect the pay.
    I am still in reserve, and I don't know how much longer we are going to stay back here. We have a training scheduled to go by while we are here. I have, at this time, 35 Americans and 4 South Koreans Soldiers.
    There was a Mullenax who came into the Company, but I didn't know him. He was from Indiana. Darling, there isn't a lot to write about here in this hole. Darling, I love you and I also trust you, for I believe you love me as I love you.
    Darling, tell all I said Hello. So good night with all the sweet dreams.

    Darling, Love for Always,
    Love Johnnie.

---

[18] *I am platoon leader doing the job of a 1st Lieutenant and only a sergeant. There 35 Americans and 4 South Koreans in this platoon*

Letter #133

Korea
April 21, 1951

To: Edith Zinkhan
Frank, WV

From: Sgt. John W. Mullenax
Co F 5th Inf Regt. APO 301
% PM San Francisco, CA

Dearest Darling,

    Well, Darling, I am back on line again. It sure was a disagreeable night last night, raining and cold, but it has warmed up today. We are pushing along slowly.

    I hope we go in tonight, reserve. They brought up hot chow this morning. It consisted of scrambled eggs, toast and coffee. We didn't get as much as I wanted, for these hills make a person hungry.

    Darling, I haven't heard from you for a few days. Hon, I sure like to get those sweet letters and to know that you are getting along OK. I often think of the Sundays we took lunch out since it had started to warm up. OK, yes, I did receive the box with the candies and cookies. It was banged up quite a bit. If you can, make the box a little more substantial because they must handle them rough.

    I got a letter from Monna and she said she was going over to Aunt Gertie's and try to drop down to see you. Well, Darling hope these few lines find you well and tell Ben I said to be a good boy and listen to you.

    Darling, I love you and counting the days to come back to you.

    Tell all Hello,
    Sweet dreams, sweetheart
    Love for always,
    Love,
    Johnnie

I am now back across the 38th Parallel Line	Julie Mullenax Van Meter

Letter #134 [19]

April 23, 1951
Korea

To: Edith Zinkhan
Frank, WV

From: Sgt John W. Mullenax
Co F 5th Inf Regt. APO 301
% PM San Francisco, CA

Dearest "Edy"

    Darling, I am sure glad to hear from you, Hon and to know that you are well and getting along OK. I am well. Darling, Thank God, for I sure had a hard night.

    Last night we fought all night and were surrounded, but we got out. "I am now back across the 38th Parallel line." We pulled back this morning. I am with our Rear Command Post and I am in the mess tent. Writing this by a candle.

    Darling, I love you and I know you love me also and I want most of all to get back to you. I am Sergeant First Class now, but I am still holding the job of a Sergeant; I led the Company out last night with my platoon. Hon, I got to get some rest tonight. I didn't get any rest last night. I guess you are hearing about this over the radio.

    Well Hon, there isn't much news. You'll get a big a picture just what is going on around here.

    Goodnight, Darling and Sweet Dreams.
    Love for always,
    Love,
    Johnnie

---

[19] *Promoted to Sgt. 1st Class. Led the Company out with his platoon, after they were surrounded. It's on the radio.*

Letter #135

April 30, 1951
Korea

To: Edith Zinkhan
Frank, WV

From: SFC John W. Mullenax
Co F 5th Inf. Regt. APO 301
% PM San Francisco, CA

Dearest Darling,

    I hope that these few lines get to you fast, for I haven't been able to write for some time. I guess you know why by the papers or news. I am at the Han River at the present time. I think we will cross over it. I have had some close calls in this retreat. Thank God I got out safe. Hon,

    I have been receiving your mail regularly. Hon, I received a box of homemade candy. It sure was good, but it is getting too warm for it now. For it melts and sticks to the paper.

    Hon, I made Sergeant 1st Class. I really don't want any more rank, only to get out of here. I am holding the job of a 1st Sgt. I was talking to one of the officers today, he asked me if I wanted 1st Sergeant. I just hope I get it, hon, for I wouldn't be on the line as much. Darling, the talk is the ERs are getting discharged in September, hope so far. I sure want to get back home. I wonder when they will start sending them back from over here.

    Darling, keep those sweet letters coming, for I watch for them. Hon, I love you and always will for you and I got along so well and we want to keep it that way.

    Goodnight and sweet dreams, Darling,
    Love for always,
    Love,
    Johnnie

Letter #136

May 1, 1951
Korea

To: Edith Zinkhan
Frank, WV

From: SFC John W. Mullenax
Co F 5th Inf Regt APO 301
%PM San Francisco, CA

Dearest Darling,

How is my Darling future by this time and how is our boy? I hope he passes this year. Tell him I said for him to study hard. Hon, I often think of the swell times we have had together. It is getting warm and it makes me think of the swell picnics we had gone out on Sundays.

I have a hard time over here, but I keep it to myself, for you have enough to worry about back there.

Hon, I only hope I get 1st Sergeant, for that would put me back with the Rear GP more. I am now doing the job of a 1st Sergeant and I don't want to be an officer for I would have to sign up for 3 years and I don't want no part of it anymore.

I am now at the Han River. I don't know how far we will go back. I have a new Company Commander the other one got wounded. He seems like a very good man. Hon, don't forget you have a man that loves you and looking forward to the day he can hold you in his arms.

Tell all hello.
Love for always,
Love,
Johnnie, Jackass.

I am now back across the 38th Parallel Line   Julie Mullenax Van Meter

Letter #137 [20]

May 2, 1951
Korea

To: Edith Mullenax
Frank, WV

From: SFC John W. Mullenax
Co F 5th Inf Regt APO 301
% PM San Francisco, CA

Dearest "Edy"

How are you by this time? Fine, I hope. Hon, I am getting along OK, as good as the condition permits. I am still a platoon leader. I have, at the present time Thirty-two men in my platoon. They wanted me to take a Battle Field Commission, but I told them no, for I would have to sign up for 3 years and I know you would not want it and I don't want no part of this Army life.

I am due to make Master Sergeant. I will take that for it is a good rank to have. Hon, I have had it pretty easy for the last 4 days and I sure needed it, for I have been pushed hard. I am on the Han River and I took my Platoon down and went in swimming.

Darling, here is a little poem. I hope you like it:

FOR YOU, MY ANGEL DEAR

I do believe that God above
Created you for me to love
He picked you out of all the rest
Because he knows I love you best
I once had a heart of mine, so true
But now it has gone from me to you.

---

[20] *Poem, written May 6.1951 still Platoon Leader with 32 men*

I am now back across the 38th Parallel Line — Julie Mullenax Van Meter

> Take good care of it, as I have done
> For you have two, and I have none.
> If I go to heaven, and you're not there
> I'll paint your face on the golden statue
> So, all the angels can know and see
> Just how much you mean to me.
> If you'd not come by judgment day
> I'll know you have gone the other way.
> So, I'll give these angels back these wings,
> Their golden halos and other things
> And just to show you what I'd do
> I'd go to "Hell" dear just for you.

Darling, I received a letter from Monna (Sister) and she said that Glen (Her Husband, Sheriff of Highland Co., VA) had Roy Eagle in jail for 10 days and let him out and had to put him in again. Hon, I also received a letter from your mother today and 2 sweet letters from my Darling.

Darling, on the boxes, don't send anything that melts, for the weather is getting hot. We got 5 cans of beer today. They were warm, so I didn't drink but a couple cans. Hot beer isn't no good.

Darling, I love you and want so much to get back to you, that is what I want; I don't want to be an officer because it doesn't mean anything to me. You are the one. Write often, Darling and don't give up hope, for I haven't.

Love for always,
Love,
Johnnie

Morning Shave with the Mirror that Edy sent him.

Letter #138

Korea
May 3, 1951

To: Edith Zinkhan
Frank, WV

From: SFC John W Mullenax
Co F 5th Inf Regt5 APO 301
% PM San Francisco, CA

Dearest Darling,

 Here goes a few lines today. Hope you are well, Hon.
 Enclosed you will find some pictures that were taken over here by a buddy. He sent them to the States and got them back, and mailed them to me. He is in another Regiment of the Division, the 19th Regiment.
 Remember the mirror you gave me. That's it in my hand in the picture. I lost it, and my shaving equipment. The chinks got our baggage truck and now all I have is the clothes I have on.
 Hon, I haven't heard from you for a few days, hope to hear soon. The last letter was dated April 23.
 Hon, I love you and I am sure you know that, for I have tried to show it. I am looking for the day to hold you in my arms.

 Sweet dreams, Darling,
 Love for always,
 Love,
 Johnnie

I am now back across the 38th Parallel Line   Julie Mullenax Van Meter

Letter #139 [21]

Korea
May 8, 1951

To: Edith Zinkhan
Frank, WV

From: SFC John W. Mullenax
Co F 5th Inf Regt. APO 301
% PM San Francisco, CA

Dearest Darling "Edy"

 How are you this time? Fine, I hope. I am getting along OK. I haven't heard from you for a few days. I have been on the hill, and it is getting hot over here. It still is cold at night.

 I had one of the men bring me this writing paper up from the rear so I could drop a line, for we had no paper with us. I am back to eating C Ratios again. I sure will be glad when I can get one of your home-cooked meals.

 I still have my platoon. I turned down the Commission. I guess I will make Master Sergeant though I have to be in grade for thirty days. I have, at present, a patrol out in front of my platoon and I have to keep in contact with them by radio, for I sent my radio man with them.

 The talk about the ERs getting out is strong over here. What do they say back there? Give me the straight dope on it if you can get it. I only hope what I have heard is true.

 Hon, the captain was going to put me in for the Service Cross for saving the Company. I got them out when we were surrounded. But my captain got wounded before he got to put me in for it. So, I guess that is all I will hear about that.

---

[21] *Captain recommending Service Cross for saving Company when surrounded by North Koreans.*

Hon, I am going on my fifth month over here. It seems like it has been years. Hon, I have lost so much weight, you won't know your old man when he gets back. Hon, I want to drop Ralph Stone a letter as soon as I get a chance. His Unit passed the other day, but I didn't see him. How is the boy? I bet he is full of mischief since it has warmed up. Well, boys are boys, Hon. You need a daughter.

How is the garden coming along? Do you have your flower garden out this year? Darling, I love you and looking forward to returning to you and hope it isn't too long. I will close for this time. Hoping to hear from you tomorrow.

Sweet dreams, Darling.
Love for always,
Love,
Johnnie

I am now back across the 38th Parallel Line			Julie Mullenax Van Meter

Letter #140 [22]

Korea
May 12, 1951

To: Edith Zinkhan
Frank WV

From: SFC John W. Mullenax
Co F 5th Inf Regt APO 301
% PM San Francisco, CA

Dearest Darling,

    I received 2 nice letters and Hon, I sure look forward to hearing from you. Darling, I also got a box of cookies, and I can't say in words how good they were. I handed them to my Company Commander and he said I will try some of your wife's cookies. I said you were my fiancé. Well, he said we know she could cook. I thought he would never get full.

    "Oh Yes" hon, I have another job. I am now Field 1st Sergeant. We got a Sergeant Replacement and he took over my job. I like this job the best and hope to keep it.

    Darling, how are you getting along? Fine, I hope. I have gotten hot chow for 2 meals for 2 days now, sure helps. Darling, I love you and can't wait for the day for me to get back to you. Hon, don't worry about your debts, for we will take care of them.

    Love for always,
    Love,
    Johnnie

---

[22] *Promoted to Field 1st Sgt. With Master/Sgt.*

Letter #141

May 27, 1951
Korea

To: Edith Zinkhan
Frank, WV

From: Master Sergeant John W. Mullenax
Co F 5th Inf Regt APO 301
% PM San Francisco, CA

Dearest Darling "Edy"

How are you by this time? Fine, I hope. Hon, I haven't written you for 24 days. I did write you last week.

We have been on the move since May 3, we are pushing ahead.

We took 570 prisoners, our Company alone. The chinks seemed glad to give up. I hope they all give up so I can come home. They are starting to send the ERs back on June 1. I don't know whether I will be on the first list.

We were praised by generals and other high-ranking officers for the good job we did. The chinks were swarming all over.

"Oh yes," I had a surprise today; we got a beer ration and a coke ration. I haven't seen a Coke for almost 5 months. Hon, this job I have now keeps me busy. I made "Master Sergeant" on May 19. I can't go any higher unless I become an officer. I don't think I want it. I have been offered it already and turned it down.

All I want is to get home, Darling, with you and get out of this hell hole. Hon, I received 2 sweet letters from you today and I also received a box of Milky Ways too. Darling, I have sure appreciated what you have done for me. You have kept my hopes built up.

Hon, when I get back, I sure will be able to eat. I have lived on C Rations almost since May 3. Darling, I miss you too and long for you, for I love you. Hon, I know you must be worrying because you haven't heard from me. It has rained here a lot and we couldn't get any writing paper, it would get wet.

Darling, don't feel harsh toward me, for I will write as often as I can when I am in reserve; I try to write every day. I hope Bennie passes this year. I will get him that gun when I come home if he does.

Well, Hon, I want to get to bed, for I didn't get much sleep last night. I was up from 12:00 pm and I am sure tired.

Darling, I hope you and Ben are getting along OK. Will close with all my Love for always.

Sweet dreams, my Darling,
Love,
Johnnie

I am now back across the 38th Parallel Line   Julie Mullenax Van Meter

Letter #142

Korea
May 28, 1951

To: Edith Zinkhan
Frank, WV

From: M. Sgt. John W. Mullenax
Co F, 5th Inf. Regt. APO 301
% PM San Francisco, CA

Dearest Darling,

    I hope these few lines find my Darling well and fine. I guess you have been working hard doing spring cleaning. I know how hard that lawn mower pushes. I don't think you should do that. For that is a man's job.

    I am reserve now. I don't know for how many days. I hope that it is a good many days, for we were on line for 24 days.

    Hon, the war news should look better, for the chinks are giving up by the thousands. If they keep it up, we should win. Hon here is a few pictures, but they aren't good.

    I received a package from you today. The case with the chocolate nuggets and peanuts. I appreciate it very much.

    Darling, you sure have done a lot for me.

    I received a letter from Geneva Fox (sister) and she said that the baby is spoiled already.

    I hear that the ERs are starting to rotate soon, I hope I go soon, but I don't look for it until July 1.

    Darling, I love you and hope to be back with you soon, for that will be a happy day.

    Sweet dreams, Darling,
    Love for always,
    Love,
    Johnnie.

Letter #143

May 30, 1951
Korea

To: Edith Zinkhan
Frank, WV

From: M.Sgt. John W. Mullenax
Co F 5th Inf Regt APO 301
% PM San Francisco, CA

Dearest "Edy"

    How are you by this time? I hope these few lines find you and Ben well.

    It sure has been a gloomy day here. It has been raining all day, and the mud is so bad the traffic has stopped.

    It sure is miserable on the guys that are in foxholes tonight. I am lucky I am with the CP and Kitchen. I am still Field First Sgt. I don't know how long I will keep the job. The "Old Man" or company commander praised me for the help I gave in running the company. I was in charge of the company a good many times when he was gone. I heard that we were going to Army Reserve. I hope the rumor is true.

    Hon, I heard that they were going to start rotating the ERs on June 3. I hope so far, that means that I will be out of here sooner than I thought. Now, this is hearsay.

    I haven't heard from you for a few days; it could be an account of the roads leading up here. Have you got the spring cleaning done yet? I bet you have been working hard. Hon, I love you and hope to see you before too long. I hope it is at least August.

    Sweet dreams, Darling.
    Love for always,
    Love,
    Johnnie

Letter #144

Korea
June 2, 1951

To: Edith Zinkhan
Frank, WV

From: M/Sgt John W. Mullenax
Co F 5th Rct. APO 301
% PM San Francisco, CA

Dearest "Edy"

Received a sweet letter from you today and I am glad to know that you and Ben are well. I hope these few lines find you well and getting along OK.

Hon, I am getting along OK; I have had it very good for the last few days, but I know that it can't keep up. I know I will have to start climbing the hills again soon.

We got twenty-two replacements a couple days ago, but we aren't up to strength yet. An order came down with the ERs on it.

In the Regiment. I am the 110th out of 300. In the company, I am the 5th on the list out of 16. I still don't know when I will start at home. I think it will be in July at least. I hope so.

We aren't what you call on the front at present. We are cleaning up the ones that are left behind. Hon, I received the box of Candy today. "Thanks a lot" I sure appreciate it very much.

Hon, here are a few pictures saved for me. Darling, I have never got so tired of a place like this. There isn't nothing here; the sooner I get out of here, the better I will feel. I am still the Field First Sgt.

The old man congratulated me for the work I had done. He told Regiment that he had to have me for Field First. We are supposed to have one.

The other captain that got wounded was going to put me in for the Distinguished Service Cross for saving the Company one night when

I am now back across the 38th Parallel Line — Julie Mullenax Van Meter

the chinks ran over us, but since he has gone to the hospital. I don't know whether I will get it or not.

Darling, How I long to be back there with you. I have had enough of this. I love you, darling, and can't wait to get back to you.

Tell all Hello,
Love for always,
Love,
Johnnie.

P.S.: *Sweet dreams, Darling.*

Letter #145

June 5, 1951
North Korea

To: Edith Zinkhan
Frank, WV

From: M/Sgt John W. Mullenax
Co F 5th Inf Regt APO 301
% PM San Francisco, CA

Dearest "Edy"

I sure hit the jackpot this time. I received 3 letters from you and 3 from Mother. I sure was glad to hear from you and to know that you were OK.

Darling, I received the box of cookies and they sure were good, only they were crushed some.

My buddies sure helped me out. I hated to ask them to save me some. I got my share, though.

I am going to the rear tomorrow to shave and wash and get clean clothes. I have been on this hill for 14 days. I hate to walk down, for it is about 3 miles. Well, Hon, about all we are doing is holding this line. Sort of waiting to hear about the Peace Treaty. We are sending out reconnaissance patrols to see what the chinks are doing.

Darling, I am hoping to rotate this month. I will write you as soon as I get the word. I know I am getting close to going and the nights and days are so much longer.

My old Company Commander is the Head of Quarters Company. He was made captain.

Hon, I was thinking of my Firecracker yesterday. I want to say I regret that I wasn't somewhere to get you a present.

All I can send you is my love and wish for you many Happy Birthdays to come. I love you and long for the day, I can see you again.

Love for always,
Love,
Johnnie

Letter #146

Korea
June 6, 1951

To: Edith Zinkhan
Frank, WV

From: M.Sgt. John W. Mullenax
Co F 5th Inf. Regt. APO 301
%PM San Francisco, CA

Dearest "Edy"

  Here comes a few lines hoping they find you well. Hon, I am in reserve now, don't know how long we will be in reserve, but hope it lasts a good while.

  They had a USO show. The Camel Caravan. I liked it very much for I hadn't seen a show for so long. It was a mixture of Hillbilly and Classical.

  Hon, I want to come home so bad. But I will have to wait. I don't know when I will get to come; I have heard so many rumors here. The RA rotation is going slow. I hope to find the straight facts about it, but everyone I ask doesn't seem to know. When I find out for sure, I will let you know.

  Mother wrote me that she was going to visit you when school is out. Hon, I received those pictures of the deer head. I thought I wrote you. I just received the picture of you and me standing by the house. It sure is a good picture. How are the new neighbors that moved into the Collins house?

  I will have to get my car out when I get back and clean it up. I bet it's all dusty and the tires are flat. I don't think they have been driving it. I may trade it in for another one.

  I just played a game of horseshoes with the company commander; I lost.

  Hon, I hope I get back before the bad weather so we can travel some and take out a picnic.

I am now back across the 38th Parallel Line   Julie Mullenax Van Meter

    Hon, I love you and long for the day to get back to you.
    I hope to hear from you today. Tell Ben I said to be a good boy.

    Love for always,
    Love,
    Johnnie
    Sweet dreams, Darling.

Letter #147

June 9, 1951
Korea

To: Edith Zinkhan
Frank, WV

From: M.Sgt. John W. Mullenax
Co F 5th Inf Regt. APO 301
% Co PM San Francisco, CA

Dearest "Edy"

How is my Darling by this time? Fine, I hope.

I bet you have a lot to do keeping the garden and house up, for I know how it is to work in that garden against the hill. Hon, I received two sweet letters from you today and was so glad to hear that you and Ben are well and that you also have your girlish figure, for I want to hold that girlish figure in my arms when I get back. Hon, I am in reserve now and hope to be here for another seven days. We sure deserve it.

When you were up at Aunt Gertie's, did she say anything about me not writing her? I wrote her some time ago, but it is so hard to write here for there isn't anything to write about, only the war, but with someone like you, which I love, I can write better.

Hon, there was a Lieutenant up to see me today to get my name. He said he heard from my old Company Commander. He didn't say what for, but I think it was for a Citation. He told some of the men that he was going to put me in for the Distinguished Service Cross before he got wounded.

Hon, I am still a Master Sergeant; that is as high as I can go as an enlisted man. They wanted to put me in for a Battlefield Commander. I said no, for I want to get out of this Army.

Hon, I haven't heard anything more about the rotation. I don't know when they will start on us ERs. Hope it's soon, for I have had enough of the damn place. I received the box of cookies and I sure liked them. They were good.

Hon, I would like to get back to your cooking and your companionship, which I miss the most.

Darling, I love you and miss you. I often think of the good times we have had together and hope and pray we can see those days again. I have to get up at 5 o'clock in the morning, so I will say good night and sweet dreams, sweetheart.

    Love for always,
    Love,
    Johnnie

Letter #148

Korea
June 12, 1951

To: Edith Zinkhan
Frank, WV

From: M/Sgt John W. Mullenax
Co F 5th Inf Regt APO 301
% PM San Francisco, CA

Dearest "Edy"

How is my darling by this time? Fine, I hope? Hon, I am getting along OK now I am in reserves. Corps Reserve. There isn't nothing here.

I am glad, though, just so we aren't online. I have been sleeping on the ground since I got over here. I guess I will sleep on the ground until I get out of Korea. I have been the Acting First Sergeant for the last 2 days. The First Sgt. Went back to the Division to check on records.

I just came back from the mess tent. I was drinking a cup of coffee with our New Company Commander. He has been here for 2 days. I think he will be alright. I let some boys go over to Seoul to see if they could pick up some extra rations. They did very well. They got some eggs, flour and seasonal things for the kitchen. They got it from the Air Force. All they had to do was tell them that they were from the 5th Infantry of the 24th Division, and they gave them some. We are short on rations, but way back, they get all they want.

Hon, I am glad to hear that you are going down to stay with Howard and Betty for you need a break. I hope you have a nice visit Hon and also tell them I said Hello.

Hon, I heard they were sending a quote down on ER rotation. I am sure sweating it out. I hope I am on it. If I am not on this one, I will be on the next one.

Darling, I wrote and told Dad to get the licenses and pay the insurance on my car. I may be getting too fast but hope not. Then I will have them when I get home.

It was 12 o'clock, and the mail clerk left for mail. Hope I have a letter tonight. Is there any of the boys returned from over here yet, that I know of. Darling, I love you and can't hardly wait to get back to you. I know then I can live like a human being. "Oh Yes," Gene Hammer is Deputy sheriff. He had a week chasing an exceptional person who locked Betty in a cell.

How is Ben getting along? I bet he has grown to be a little man by this time. I'm glad to hear that he passed this year. Well, darling, I hope it isn't too long before I get to see you. I will close for this time.

Sweet Dreams, Darling. I love you. Love for always,

Love,
Johnnie

I am now back across the 38th Parallel Line	Julie Mullenax Van Meter

Letter #149

June 12, 1951
North Korea

To: Edith Zinkhan
Frank, WV

From: M/Sgt John W. Mullenax
Co F 5th Inf Regt APO 301
% PM San Francisco, CA

Dearest "Edy"

Darling, here goes a few lines of the letter. You know I am well and getting along OK and thinking of you every day. Hon, I received 2 lovely letters and I am so glad to know that you are well.

Hon, I am still on the same hill. I have been here in this position for 19 days. I was down yesterday and stayed until this morning. I saw Jack Bennie, Errol Flinn and 4 girls. It sure was good to see American girls again. It was a good show. Jack looked a lot older since the last time I saw him in Europe.

Hon, I have been sweating out a rotation list down. There are rumors, how I hope they are true, for I am next on the ER list to go. Hon, we are hoping that this Peace goes through.

We have the war news on now. It doesn't sound like they have gotten anywhere as yet. I did get to take a bath and get clean clothes yesterday when I was down. The old man and me stayed down on the line, but the others had to come up last night.

It has been raining a lot here the last few days, but this is a nice evening. Tell Howard I sure would like to have been there when they were there.

Hon, I love you and miss you so much that words can't express how much.

Tell Ben I said to be good and help Mother in the garden. Hon, I will write you as soon as I get word for me to go home.

Tell all "Hello"

I am now back across the 38th Parallel Line   Julie Mullenax Van Meter

    Love for always,
    Love,
    Johnnie.

    P.S.: *Sweet dreams, Love, Johnnie*

I am now back across the 38th Parallel Line   Julie Mullenax Van Meter

Letter #150

June 15, 1951
Korea

To: Edith Zinkhan
Frank, WV

From: M/Sgt John W. Mullenax
Co F, 5th Inf Regt. APO 301
% PM San Francisco, CA

Dearest Darling,

    Received 2 sweet letters from you yesterday and I am glad to know that you are well and enjoying yourself down at Betty and Howards. Hon, I don't see why you don't stay longer for you need the break. I also received the box of Welch's Candy you sent me. Thanks a lot, hon. Darling, we are still in reserve and getting plenty of candy, so don't send any more packages. But keep the letters coming.

    Hon, they have started the ER's rotating 2 from our Company and are leaving July 18. I will be the fourth on the list after they leave.

    I will sure be happy when they say I can leave here on my way home.

    Darling, I sure would like to get home before the cold weather sets in. They asked me if I wanted to be an officer again. I said no, I wanted to go home. My captain left this morning. I hated to see him go.

    I don't know how our new captain will be.

    I am still the Field First Sgt. I have a very good job. Hon, I will be glad if I can write you and say I'm coming home. I will let you know just as soon as I do.

    I told Dad to buy licenses for my car, so I will have them as soon as I get home. I was over and talked to the old man last night before he left and helped him drink his whisky. It sure seemed strange to get a drink over here.

Our chow has improved some since we are back here in Corps Reserve. I guess I will be wanting to eat all the time when I get home.

Darling, I love you and often think of the swell times we have had together. Tell Ben I said "Hello" and be a good boy.

Love for always,
Love,
Johnnie

I am now back across the 38th Parallel Line   Julie Mullenax Van Meter

Letter #151

Korea
June 20, 1951

To: Edith Zinkhan
Frank, WV

From: M/Sgt John Mullenax
Co F, 5th Inf Regt. APO 301
% PM San Francisco, CA

Dearest "Edy"

    Received 2 letters from you, plus the joint letters and a letter from Aunt Gertie. I am so glad to hear from you and to know that you are well and getting along OK.

    I am also glad to hear that Mother is enjoying herself on her visit. I am glad that you enjoyed yourself on your visit to Betty's.

    Hon, I am not doing no more than I have to do. I don't think my New Company Commander likes it much, but I think the boys need a rest by being fighting so long.

    I just found the Company; I sent 2 platoons out to put up barbed wire and a platoon to fire their weapons. So, I slipped out under a tree to write you a letter.

    One good thing is that we have the shower unit set up. Back of our Company along the river. I heard yesterday that we would be in reserve for twenty more days if the Reds don't counterattack. I hope we do stay that long. I hope and pray that I am out of here before I have to go back online.

    "oh," I have a headache today. They are breaking PX rations down. I got them last night at Battian and the mail. I didn't get back until 12:30 last night. I made the mail clerk break the mail down. I knew I was going to get a letter from you.

    Hon, don't send any more boxes. I may not receive them but keep writing me until I tell you I have started home. Hon, these days seem so long waiting to go home. There are 2 ERs ahead of me yet.

Ben sent me a report on the sawmill. It sounded like they were doing very well.

Well, Darling, I had better get busy. I love you, Darling, and can't wait for the day when we can see other.

Love you always,
Loving you only,
Love,
John

I am now back across the 38th Parallel Line	Julie Mullenax Van Meter

Letter #152

Korea
June 21, 1951

To: Edith Zinkhan
Frank, WV

From: M/Sgt John W. Mullenax
Co F 5th Inf. Regt. APO 301
% PM San Francisco, CA

Dearest "Edy"

How are you by this time? Fine, I hope. I received a letter from you today. So glad to hear from you and to know that you are well and getting along OK. Hon, I am glad to hear that Mother is enjoying her visit.

Hon, I am moving up on line tomorrow to relieve the 3rd Battalion of the 7th Division. Well, I can say that I enjoyed our reserve for what time we were off the line. I am sitting in a Jeep writing this. The men are loading up the truck. We move up about sixty miles tomorrow.

We have a new company commander. I don't know how well he will handle the Company.

I just got back from taking a shower and a clean change of clothes. This is such a beautiful evening that I would like to be home with you. Hon, I often think of the swell evenings we spent together. I just hope I get back before the cold weather.

Darling, I am going to take a vacation when I get back.

I was just told that I would get up at 3:30am in the morning, so I am going to have to get my things together. So, I will close for this time.

Love for Always,
Love, Johnnie

P.S.: Sweet Dreams, Darling.

Letter #153

Korea
June 25, 1951

To: Edith Zinkhan
Frank, WV

From: M/Sgt John W. Mullenax
Co F 5th Inf Regt. APO 301
% PM San Francisco, CA

Dearest Darling,

    How is my Darling by this time? Fine, I hope. Hon, I am getting along OK, only I am very dusty, you can tell by the letter being messed up.
    Darling, I am on the hill again. I came back on June 21. I'm still the first field sergeant. Some of the duties are to see that we get rations and ammo and see that the wounded get off the hill. I have 2 men working with about 60 civilians who carry this on the hill for us.
    One of our big problems is water. It is very hot over here and so far off the hill to the water.
    Darling, how are you getting along with the house and garden. I guess our little man is getting big enough to help a lot.
    I read in the Stars and Stripes (Magazine) that the World War II veterans will get out first. There are only 3 in this Company, including myself. I hope we do.

    Love for always,
    Love,
    Johnnie, J. A.

Letter #154

Korea
June 25, 1951

To: Edith Zinkhan
Frank, WV

From: M/Sgt John W. Mullenax
Co F 5th Inf Regt. APO 301
% PM San Francisco, CA

Dearest "Edy"
    Hope these few lines find you well and getting along fine. I am getting along OK at present. I haven't been doing very much.
    We are on a high hill holding a line. We are putting up a defensive line. Here this makes the second time I have been across the 38th Parallel line. There are some GIs out in front of us that makes it a lot safer here.
    I have received a good many sweet letters from you while I have been back on the hill. I haven't gotten to write you too often since I have been up here. I borrowed this paper from the Company Commander to write you.
    Hon, I wrote you some time ago not to send me any more boxes. I am getting plenty of candy etc., now. I sure got a surprise last evening. I got ice cream, and it melted a lot, for it took 4 hours to get it up here.
    Mother wrote and said that she enjoyed her visit with you. Hon, I will be glad to get something fresh from our garden, for I am getting tired of canned food.
    You said I would miss my chicken. Well, I will get my chicken when I get back. "Ha" you.
    Darling, I only hope to get to leave this month. Hon, I sure like to hear you say, "Johnnie, breakfast is ready". We are supposed to get some beer up here today. Darling, I love you and hope it isn't so long before I see you.
    Love for always,
    Love,
    Johnnie, J. A.

I am now back across the 38th Parallel Line   Julie Mullenax Van Meter

Letter #155

Korea
July 3, 1951

To: Edith Zinkhan
Frank, WV

From: M/Sgt John W. Mullenax
Co F 5th Inf. Regt, APO 301
%PM San Francisco, CA

Dearest "Edy"

    I have received a sweet letter today, and I am sure glad to hear from you and to know that you are well.

    I am still on the hill; I just got through giving out C rations and 2 cans of beer per man.

    We have one ER leaving on the 10th, and I am on the next one to go. I don't know, but it will be after the 10th of the month. I hate to think about spending the 4th on this hill. I guess it will be a quiet 4th unless the chinks hit us.

    We have a radio rigged up here on the hill. A 310 and 619 sitting together. One transmitting and the other receiving. We have been getting the news about the Peace Treaty. I hope it proves out OK. I get to go down to the rear day after tomorrow to get a bath and a shave. We have been sending 6 men down a day. The old man said that one of us had to be here at all times.

    You said that lawn mowing would help your figure. I sure would like to see your figure and hold you in my arms. Hon, for it, it seems as though it has been 4 years.

    I think Dad is getting the license for my car, so I will have them when I get home.

    You said that your mother was there. What is Pits doing, teaching this fall in Greenbrier County? How is the stove working now? I guess the radio went on the blink, for it is so old.

I am now back across the 38th Parallel Line			Julie Mullenax Van Meter

    Hon, everyone has such high spirits about the war, being over if it isn't, the morale of the men will hit the bottom.
    Darling, I love you and hope to be with you before too many months. You and I will have a good many places to go when I get back. I will close loving you more each day.

    Love for always,
    Love,
    John.

I am now back across the 38th Parallel Line	Julie Mullenax Van Meter

Letter #156

North Korea
July 7, 1951

To: Edith
Frank, WV

From: M/Sgt John W. Mullenax
Co F 5th Inf Regt APO 103
% PM San Francisco, CA

Dearest "Edy"

    Received your letter of June 28; so glad to hear from you and to know that you are well and getting along OK. Hon, I know that you are expecting me home soon. I also thought that. I would be on my way before now, but it seems that they aren't going very fast. We have one ER to go on the 10th of this month, and then I am next on the list. I heard a rumor that there would be a list down again on the 15th of this month. I hope it's time, darling, for I sure want to get out of here. You can't imagine how the days drag along when I know that I am next.

    Hon, I am down off the hill at the rear CP. I came down yesterday morning, and I am going back up this evening. I got a shower and a clean set of clothes. I feel good to get cleaned up after going so long. We also had some beer here yesterday. We got some spring water to cool it. My old company commander came over and had a few yesterday with us. He is one swell guy.

    Darling, tomorrow is the day we have been hoping for, the 8th. If it falls through, there will be a disappointed bunch of men, for we are hoping that they will agree on a cease-fire.

    Darling, how is the garden coming? Has Hoover's cow been in it? How long does your Mother plan to stay with you?

    Darling, I love you and hope I get to see you soon, for I have missed you so much. I often think of the things we did and the swell times we had together. Darling, don't send any more boxes, for they may not get there; if I should leave, I will close, loving you and only you.

    Love for always,
    Love, Johnnie

I am now back across the 38th Parallel Line   Julie Mullenax Van Meter

Letter #157 [23]

North Korea
July 13, 1951

To: Edith Zinkhan
Frank, WV

From: M/Sgt John W. Mullenax
Co F 5th Inf. Regt. APO 103
% PM San Francisco, CA

Dearest "Edy"

 Will drop you a few lines hoping that they find you well. I am getting OK, Hon, but I am getting so damn tired of this hill. I have been on it for 21 days. But I guess it is better than pushing forward.

 Hon, I know you are disappointed that I haven't started for home. Well, I am too. They tell me this is supposed to be a drop for so many to go home, but it hasn't come down as yet.

 I hope it comes down soon for hate wanting to go. Darling, I love you and hope that I can get home before the cold weather. How are your neighbors by now? I guess Floyd is still running the filling station.

 Mother said they had about all the timber sawed that we had bought. I don't know whether I will stay in that business when I get back or not. What do you think about it? Tell Ben I would like to know how many fish he caught, and no fish tales.

 Hon, I am making some hot coffee on a little gas stove. I will be glad to get a cup of the coffee you make. Hon, you said something about these Korean girls. Well, I don't get to see many, and I don't like the looks of them, and these people are dirty so don't worry about me falling for on go these chinks.

 Honey, you are the one I want to see, and I am about nuts waiting. I will close for time loving only you.

   Love for always,
   Love,
   Johnnie

---

[23] *Been on Front line for 21 days.*

Letter #158

Frank, WV
July 14, 1951

To: M/St John W. Mullenax
Co F., 5th Inf. Regt. APO 301
% PM San Francisco, CA

From: Edith Zinkhan
Frank, WV

Dearest Johnnie,

    A few lines and hope they will find you OK. I have received only 1 letter this week. We are fine, Hon, and Ben is really getting big.

    Esta was up for a while last night. Oh yes, Jen and Ralph are expecting another baby. They're going in for a big family, don't you think? Ha! And still not housekeeping. I mowed the yard last evening, and Mom made some raspberry jelly.

    I just got up, and I am still half asleep. There were 2 or 3 loads of soldiers that came in this week; there were 2 Reservists. I keep hoping you're on some of them. Boy, I hope you're home soon and before it gets cold again.

    Hon, there isn't much to write about. I love you, sweetheart, and I have missed you so darn much, and I sure will be happy when you're home again and when we will be drinking coffee together. Coffee is my breakfast. Coffee and cigarettes.

    Well, sweetheart, I'll sign off for now and hope I get a letter today.

    All my love, for always,
    Your Battleax "Edy"

Letters #159

Frank, WV
July 16, 1951

To: M/Sgt John W. Mullenax
Co F 5th Inf. Regt. APO 301
%PM San Francisco, CA

From: Edith Zinkhan
Frank, WV

Dearest Johnnie,

    A few lines so early in the morning. I hope they will find you well.

    This weekend went pretty fast. I worked up at your Aunt Gertie Saturday night and Sunday afternoon and night. They were pretty busy. She said to tell you they were fine. George has a pickup now. I have 15 glasses of jelly made so far. Current and Raspberry.

    Hon, I didn't write last night; I was pretty tired. Their girl was on her vacation, and she will be back today. Boy, I sure would love to have a vacation, you and I. I hope you're home soon. I have been dreaming about you lots. It has been really hot here. Guess Mom will leave this week, but she is coming back, I think.

    So, I may not get to go over to your home after all, guess Ben will go.

    Sweetheart, I love you very, very much and hope you're home real soon. Hon, I am going to sign off for now.

    With all my love for.
    Always,
    Love,
    "Edy".

    P.S.: *Floyd and Ruth went to Lewisburg yesterday on a picnic with their daughter-in-law and grandchildren. Sam and Ruth Jennings have finally got moved.*

Letter #160

Frank, WV
July 16, 1951

To: M/Sgt John W. Mullenax
Co F 5th Inf Regt. APO 301
%PM San Francisco, CA

From: Edith Zinkhan
Frank, WV

Dearest Johnnie,

    I also hit the jackpot today. I received 3 letters from you, and it sure was good to get them. Today is the 16th, Hon, and I sure hope you're on your way home.

    I have been doing a little bit of everything today. Washed the laundry. Mom and I picked almost a gallon of Raspberries, and I hoed the corn. The garden looks pretty good. My peas will be ready by next week to can. Hoover's cow hasn't got hungry yet. I took Ben and a dog of the other's kids over to the dam this afternoon.

    Pits put his application for Pocahontas County, WV. But he doesn't know where he will teach. The stove is working fine. When I get something to cook on it. Prices are sky high, Hon, and still going up. I know what you mean, hon, by the days dragging by. I hope you're on your way. You will never know how happy I will be when you're home again.

    Sweetheart, your love was the first birthday present a girl could get. I hope you will always have your love, Hon. And always be your firecracker. I haven't exploded for a long time. Only when Ben gets me going.

    Glad you like your cookies; they didn't turn out too good. Sweetheart, I love you with all my heart, and I will say goodnight, sweetheart, and sweet dreams.

    All my love for always,
    Love,
    Your Battleax.

Letter #161 [24]

North Korea
July 16, 1951

To: Edith Zinkhan
Frank, WV

From: M/Sgt John W. Mullenax
Co F 5th Inf Regt APO 103
% PM San Francisco, CA

Dearest "Edy"

    Darling, I received good news last night. I am to start home on the 26th of this month. I will leave the Company on that date. I don't know how long it will take me to get to the States. There is another man taking over my job. I will teach him the routine, and I think the Old Man will let me go back to the rear CP until I rotate.

    Hon, it sure is a great feeling to know I am starting home soon. Hon, you can stop your letter writing if you want to, for it won't catch up with me when I start for the States. I will keep writing you letters. You know how I am getting along and the progress I am making toward home.

    Darling, I hope these few lines find you and Ben well. Hon, I love you, and you are always on my mind. I sure will be glad to get away from this cooking and C. Rations. I yearn for some of your good cooking.

    I told the Old Man what a good cook you were. He's said he knew by the cookies you were a good cook.

    I am going to try to bring my helmet home with me, for it has a hole in it. I think they will let me bring it. I hope so.

    The company commander just told me we were getting another Lieutenant. He is going back to Division Rear today for about 3 days. It has been raining here for a few days, and it stays foggy on top of this mountain. Well, Hon, I will close by loving only you and hope it isn't too long before I see you.

    Darling, for I love always,
Love, Johnnie
P.S.: *Sweet Dreams*

---

[24] *Should start home on the 26th of July, 1951*

Letter #162

North Korea
July 18, 1951

To: Edith Zinkhan
Frank, WV

From: M/Sgt John W. Mullenax
Co F 5th Inf. Regt APO 103
% PM San Francisco, CA

Dearest "Edy"
    Received 2 sweet letters from you today and I am so glad to hear from you and to know that you are well. I am well, Hon, but I am still on the hill.
    The company commander left yesterday to go back below Seoul, and he asked me to stay on the hill until he returned. When he gets back, I guess I will get to go to the rear until I start home. At least, I hope so.
    Hon, I have been on this hill for 24 days. It sure will be good to get to sleep in a bed again. I have slept on the ground for 6 months.
    Hon, I sure will be glad to hear you calling Johnnie again, for that is music in my ears.
    I just was interrupted by a one-star general who was here looking over our positions. Hon, I can't hardly wait until the 26th to start on my way. Hon, your letters won't catch up with me when I start home.
    I love you, Hon, and I hope to see you in the near future.

    Love for Always,
    Love Johnnie.

Letter #163

Frank, WV
July 18, 1951

To: M/Sgt John W. Mullenax
Co F 5th INF Regt. APO 301
%PM San Francisco, CA

From: Edith Zinkhan
Frank, WV

Dearest Johnnie,

    A few lines for tonight. Hope you're well and on your way home. Mom went home today at Big Run. Guess she will be back Saturday or Monday. I made more jelly today. I have 23 jars in all. That's about all I did. Sweetheart, I am so darn glad you will be home soon. These 8 months have been years for me. I hope I don't show it. I think I am gaining weight. I hope not, for I know you don't like fat women. Ha.

    Received a letter today from Mrs. Zinkhan. She said that Betty had gotten married. We're well, Hon, and Ben is mean as ever. There isn't much to write about. It has been very hot here.

    Paul Collins will be home for a 15-day vacation this weekend. He is taking his family back to Honolulu. Sweetheart, I don't know what to write. I love you very much, and I will be happy when you're home again.

    The kids were listening to the fight, so I had to stop writing. Joe Wilcote and Charles. Joe won in the 3rd round. So, it didn't last long.

    Hon, I'm going to say good night, Sweetheart and Sweet Dreams.

    With All My Love for Always,
    Love Battleax.

Letter #164

Korea
July 21, 1951

From: M/Sgt John W. Mullenax
Co F 5th Inf Regt APO 103
%PM San Francisco, CA

To: Ms. Edith Zinkhan
Frank, WV

Dearest "Edy"

 How are you by this time? Fine, I hope. Hon, I received a sweet letter from you yesterday. Sure, are glad to hear from you, Darling.

 Hon, I am off the hill at the rear CP waiting until I rotate on July 26. It sure is a great feeling to get off the hill for good and to know that I don't have to go back up there. I think it will take me about one month to get home after I start.

 I am going to try and bring my helmet home. It has a bullet hole in it. One of my close shaves. I was put in for the DSC. It may be 6 months before I get it. Hon, it sure has been miserable here. It has rained for three days, and the mud is deep.

 But I can take anything after I am off the hill and not eating C Rations. Darling, I am longing for one of your cooked meals. Sometimes I can hear you in my sleep calling me to breakfast like you used to do.

 I also received a letter from Mom, and she said that she had received a nice letter from you. She said they had the lumber yard full of timber and had shut down to make hay. How is Ben been behaving? Tell him I hope to see him soon and we will go to the movies.

 Darling, I will close, leaving only you and thinking of you each day.

Love for Always,
Love,
Johnnie

I am now back across the 38th Parallel Line Julie Mullenax Van Meter

Letter #165

July 21, 1951

This Letter was Returned to his Home Address of Blue Grass, VA

To: M/Sgt John W. Mullenax

From: Edith Zinkhan
Frank, WV

Dearest Johnnie,
    Here goes a few lines. Hoping you're OK and well. Everything here is dead as ever. Mom came back yesterday, but she is going back today, and guess Ben will go over to your place. I was supposed to go, but the peas are ready to can. What they are, and the raspberries are ripe. So, I guess I will can them all by myself.
    I hope you're on your way home, I don't want to be getting too many letters from you lately, and it's hard to write when I don't get any. The paper stated they're releasing 90,000 reservists within the next 3 months. Sweetheart, I'll be really glad when you get home. I made some kitchen curtains yesterday.
    Sweetheart, I don't know of anything to write, only I love you with all my heart and waiting just for you. These days are really dragging by.
    Well, Hon, I'll close for now.

    All my love for Always,
    Your Battleax,
    "Edy"

I am now back across the 38th Parallel Line   Julie Mullenax Van Meter

Letter #166

July 22, 1951

To: M/sgt John W. Mullenax
Co. F 5th Regt APO 103
%PM San Francisco, CA
This letter was returned to Blue Grass, VA

From: Edith Zinkhan
Frank, WV

Dearest Johnnie,

 This sure has been a long blue Sunday for me and really hot. Hope these few lines will find you on your way home and well. I'm all by my lonesome today. Mom left yesterday, and Mildred and Joe came over last night and took Ben home with them. I went to the movies. They had all the children with them except the baby.

 I canned 3 pints of peas today. I have been walking around in circles. I hope there aren't many more Sundays like today. There was a load of soldiers that came in today. I'm hoping you're among them, but probably no good luck. Ruth and Floyd went to Washington after Paul yesterday. I see they have gotten back.

 Sweetheart, I love you. Hope you haven't changed your plans for us. You haven't mentioned them for a long time. I have thought a lot about them, for I have had plenty of time to think. Maybe I do too much thinking.

 Sweetheart, I'm running out of news, so I'll sign off for now.

With All My Love for Always.

Your Battleax,
Love "Edy"

I am now back across the 38th Parallel Line   Julie Mullenax Van Meter

Letter #167

July 23, 1951
Frank, WV, USA.

To: M/Sgt John W. Mullenax
This letter was sent home to Blue Grass, VA
Co F 55th Ing Regt APO 103
PM San Francisco, CA

From: Edith Zinkhan
Frank, WV

Dearest Johnnie,

    I received a sweet letter today from you. Hon, you wrote on July 12. I hope you're on your way home by now. Today has been another long day for me. I have done a little bit of everything; made 12 jars of raspberry jam and jelly and washed clothes.

    We had a nice shower late this afternoon. We really needed it. I bet it did make you feel good to see Jack Benny's show and to see a couple of American girls. Just so you don't go falling for them. Ha. Remember, you have a Battleax back home. Only kidding. Hon., I will stay by my lonesome for tonight. So, I don't guess anything is going to get me. Ha, I sure do miss Ben, even if he worries me a lot.

    Everything is dead as ever. Hon, I miss you terribly and love you dearly. I'll sure be glad when you're home again. I know my letters are dull and not very interesting, Hon, I don't go anywhere, and if I do, I don't enjoy myself. So, I would rather stay home.

    Hon, I am going to say good night and sweet dreams, sweetheart,

    All my love for always,
    Love "Edy"

I am now back across the 38th Parallel Line Julie Mullenax Van Meter

Letter #168 [25]

Korea
July 25, 1951

To: Edith Zinkhan
Frank, WV

From: M/Sgt John W. Mullenax
Co F. 5th Inf. Regt. APO 301
% PM San Francisco, CA

Dearest "Edy"
    How is my Darling by this time? I received a nice letter from you yesterday. So glad to hear from you, "Hon". Darling, I am still waiting and how the days and hours go slowly. I leave tomorrow at 6 o'clock in the evening. I am not doing a thing now, just eating and sleeping. I went to a movie last night in the rice paddy. Fat Mam was the name of it. It was a mystery. Hon, I lay awake at night thinking of you, longing for you.
    Hon, I leave here and go back to Service Company and stay one night, then I go about 18 miles on the other side of Seoul. I board a ship from Incheon, Korea, to Japan. I only hope I fly back, but they aren't in any hurry to get me back as they were to get me over here.
    You said that you helped Aunt Gertie. I am glad to hear that they are doing fine. How is Ben? I bet he is a big man now. Does he help you with the work? Tell him I said Hello and be a good boy and listen to Mother.
    When a person gets his orders to ship out, he gets to come down to the rear CP five days before he leaves. Well, when I got my orders, the Old Man had to go to Seoul to get an examination for Regular Army. He told me, or rather asked me if I would stay until he returned in two or three days. Well, I stayed for three days, and a Lieutenant said he may not be back for five days or a week. So, I came off the hill. He was pissed off when

---

[25] *Received orders to ship out*

he got back. He thought the Company couldn't run without me. I told him they had just well get used to being without me now.

I don't give a hoot what they do from here on out. I just want to get out of here. Hon, do you think I should get a new car? I have the money saved, of course, the one we have doesn't have any miles on it. I found out that I could buy one in Tokyo, Japan and save the tax by getting a sales slip, then picking it up at the factory. I hope I can get to Tokyo, Japan and check on this.

Mom said they had a yard full of lumber. I will have to check on the business, but first, you and I have a vacation, just us, than the business. Hon, I know what you have gone through with staying at home waiting for me. I hope I can repay the long hours you have spent by yourself. Hon. Keep the home fires burning. I hope to be home before too long. I don't know what this peace settlement will bring. I just hope that it doesn't change my time here. Hon, I love you and yearn to hold you in my arms and tell you how much I love you. I will write you at every Station on my way back.

Loving only you,
Love for Always, Johnnie

I am now back across the 38th Parallel Line   Julie Mullenax Van Meter

Letter #169

This letter was returned to Blue Grass, VA

July 25, 1951
Wednesday night,

To: M/Sgt John W Mullenax
Co. F 55th Inf. Regt APO 103
% PM San Francisco, CA

From: Edith Zinkhan
Frank, WV

My Dearest Johnnie,
    Here goes a few lines hoping they will find you well and OK. I am well, I guess. But it's terrible being alone. Guess Ben is having a time over at your home. I picked more raspberries, and I am canning peas now. I really miss Ben, guess he will be home Saturday, I hope. I have only received one letter this week. I sure hope you're on your way home.
    Esta was up for a while this morning. Everything is dead, as always. Hon, Sweetheart, I don't know of anything to write. So, I am going to make this letter short. I love you, Sweetheart, and will be so glad to have you home again.
    Goodnight, sweetheart and sweet dreams.

    All my love,
    For always
    Love "Edy"

CHAPTER SEVEN

ORDERS TO SHIP BACK TO THE US

Letter 170 to Letter 175

Letter #170 [26]

South Korea
July 28th 1951

To: Edith Zinkhan
Frank, WV

From: M/Sgt John W. Mullenax
Co. F 5th Inf. Regt. APO 103
%PM San Francisco, CA

Dearest "Edy"
    Hon, I am writing a few lines to let you know that I am on the way. I am, at the present time, close to Seoul at a Processing Center for Rotation.
    Darling, I leave here on July 30 to Incheon Harbor, Korea, there I will go to Japan. We have been here processing. I have checked my records to see if they are up to date, which they are. Otherwise, I don't do nothing. I am in charge of a picket which consists of twenty men.
    Darling, I hope it doesn't take so long for I long to see you. It seems as though it has been years.
    I will close loving only you and hope to see you before too long. I will write from Japan.

    Love for always, my darling
    Love,
    Johnnie.

---

[26] *At Processing Center near Seoul waiting for rotation*

I am now back across the 38th Parallel Line   Julie Mullenax Van Meter

Letter # 171

This letter was returned to Blue Grass, VA

July 28, 1951
Saturday

To: M/Sgt John W. Mullenax
Co F, 5th Inf. Regt. APO 301
%PM San Francisco, CA

Dearest Johnnie,
    Here goes a few lines hoping you're well and not in Korea. I don't seem to be getting much mail from you these days. I know it's hard to write often. I have been writing every other day. I'm hoping you're on your way home.
    Well, Hon, I am still here; nothing has gotten me yet. I hope Ben gets home today. This sure has been a long week for me. Elsie was over for a while last night and Kathyrn Davis.
    I was down to Durbin the other evening for about 5 minutes, and I saw Ben going down with a load of lumber and Brooks.
    I had two dresses made. They look nice, but one is too big. Sweetheart, it will be so darn good to have you home again, and I hope it's for good. Sweetheart, I love you with all my heart. Maybe you get tired of reading those words. I hope not.
    Hon, I'm running out of something to write about. So, I'll sign off for now

    With All My Love For

    Always Love,
    "Edy" or Battleax

Letter #172

July 31, 1951
Frank, WV

To: Benjamin C. Zinkhan, III
% Walter Mullenax
Blue Grass, VA

From: Edith Zinkhan
Frank, WV

Dear Bennie,
    Received the card Mildred wrote and glad you are having a nice time. I have missed you lots. The Lamberts' children wouldn't stay. So, I have been staying alone. So, hurry home, for it's lonesome here without you. Johnnie is supposed to be on his way home. He thought he would leave on the 26th.
    Your dog Skipper is fine and barks a lot. Tell all Hello for me and you be a good boy and listen to them.

    Hope to see you soon,
    Love Mother.

    P.S.: If you don't come home until Saturday, I am going to Elkins with Elsie, and if I am not here when you come, go over to Mrs. Lambert.

I am now back across the 38th Parallel Line   Julie Mullenax Van Meter

Letter #173

July 30, 1951

To: M/Sgt. John W. Mullenax
Co. F 5th Inf. Regt. APO 103
%PM San Francisco, CA

From: Monna Mullenax Hammer (Sister)
Monterey, VA

Dearest John,

    Sorry I haven't got around to writing to you more often. I was so in hopes you would be starting home around the 15th of July. Do hope you will be starting soon. Praying you are well and will be OK. I spent the day over at Aunt Gertie's yesterday. She is still not crazy about the place. Said Edith Mullenax (sister) had been up and helped 2 days and was good help. Aunt Gertie is coming over here this week for 2 or 3 days. I wanted to go to Staunton to do some shopping. They bought a secondhand pickup truck.

    I saw Mother last week at Bobby Simmons' Funeral. Eagles sent Ray to the Western State Hospital for several weeks. He is back now and has been doing pretty good so far.

    I just didn't send The Recorders here of late. Thought you would be home before they got there. I am putting some in the mail today any way.

    I haven't been getting many cars. They are getting scarce. Haven't got the car fixed yet, that Gene wrecked. Went to Mary Hammer's church wedding in Waynesboro, VA, couple weeks ago. It was a pretty wedding. Dorothy, her husband and little girl was there. Will try to write to you more often. Hope we will be seeing you before long.

    Lots of Love,
    Monna

I am now back across the 38th Parallel Line					Julie Mullenax Van Meter

Letter #174 [27]

August 9, 1951

To: Edith Zinkhan
Frank, WV

From: M/Sgt John W. Mullenax
Sasaki, Japan

Dearest Darling,

    How are you by this time? Fine, I hope. Hon, I am OK, only I am off at this rotation. It took me 12 days to get here from Korea. Hon, I don't know when I will ship out of here. I hope it is soon. How it is not here! I have been what you might call a Prisoner in Service. I left the Company, and they don't let us out of these coups. Some of the boys are flying from here. I hope I do, also. I will make it home quicker.

    Hon, excuse the letter; I can hardly write without perspiring over it. Darling, I am so anxious to get home to you. I get so bored in the camps, for they sure could send us out faster. There are about 10,000 soldiers here, and I have to stand in line everywhere I go.

    Hon, I love you and hope to hold you in my arms soon.
    I will write when I get to the West Coast.

Love for always,
Love, Johnnie

---

[27] *Arrived at Sasaki, Japan Shipping out on the USNS General Simon B. Buckner*

USNS General Simon B. Buckner, the ship on which John returned home.

I am now back across the 38th Parallel Line   Julie Mullenax Van Meter

Telegram #175 [28]

The final letter is a Western Union Telegram

August 30, 1951 @ 8:47 am
Fort George G. Meade, MD
Ms. Edith Zinkhan
Frank, WV

Arrived in Fort Meade Tuesday. Hope to see you Saturday or Sunday.
I am waiting for my release.

Love,
John.

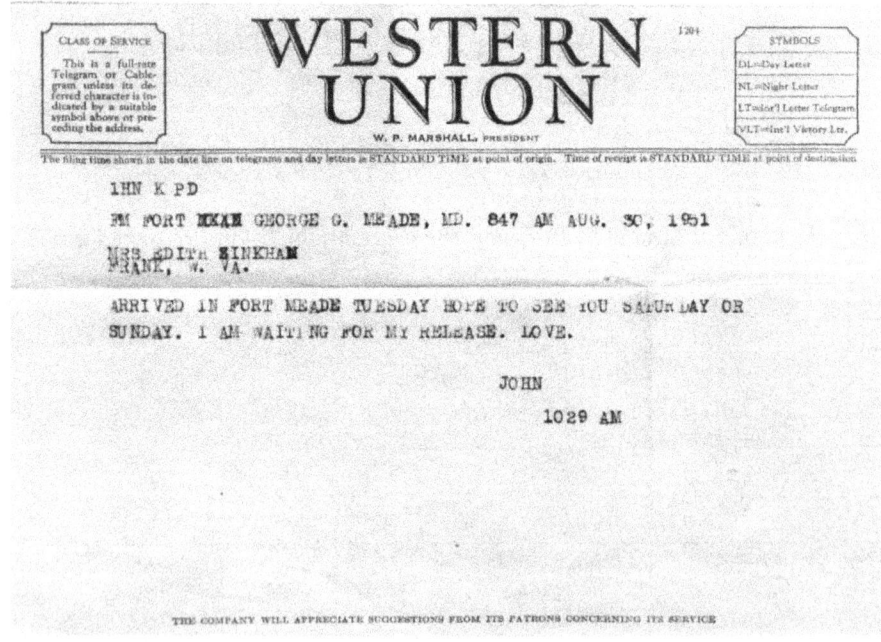

Final communication from M/Sgt John W. Mullenax to Edith Zinkhan.

---

[28] *Final Letter, a Telegram from Fort George G. Meade, MD*

CHAPTER EIGHT

HOME LIFE AFTER KOREA

On August 28, 1951, M/SGT Mullenax received word from the Department of the Army that he had received a Bronze Star for his exemplary actions in Germany on November 28, 1944. Three days later, he was discharged from active duty.

He returned to Blue Grass, Virginia and began working with his father at Mullenax Lumber Company. Trees were cut, lumber was produced and the business strived.

August 7, 1953 John and Edy were married in Rockingham County Virginia at the Court House. They settled in a farmhouse near Blue Grass, Va. Mullenax Lumber Company was moved to the additional acres purchased with the house.

August, 13 1958, John and Edy welcomed home a baby girl, Julie Annette Mullenax. With Edy's son, Ben the family was complete.

As Mullenax Lumber Company grew, it was decided that there was a need to move to a larger space. June 15, 1965, John purchased a sawmill that was going out of business. It was located in Franklin, West Virginia, 25 miles north of Blue Grass, Va. After the business moved, the family didn't relocate until 1969. Mullenax Lumber Company combined with Franklin Industrial Company was renamed and Franklin Enterprises, Inc. was formed. During the lumber boom of the 70's, our lumber was shipped to North Carolina, Virginia and other states to be made into furniture. John opened a building supply store in 1971 to aid the people of Pendleton County to be able to build or repair their homes. Also, he opened a furniture store to sell the furniture that was being made in North Carolina.

John was a member of many Veterans' Organizations. Veterans of Foreign Wars, American Legion, Order of the Purple Heart. He was instrumental in creating the Disabled American Veterans Organization for Pendleton County. He was also active in the Ruritans, and the Democratic Party for Highland County Virginia. He was a member of the Pendleton County Lion's Club and Franklin #769 Loyal of the Moose, in Pendleton County.

It would take twenty-five years for the Army to recognize John for his courageous leadership at Pisi-gol. He finally received his second Bronze Star for Valor.

A private ceremony was conducted in 1977, and the National President of the Veteran of Foreign Wars, Richard Homan, presented M/SGT Mullenax with his Korean Bronze Star.

When asked if he would do it all again, John said, "I wouldn't take anything for the experience I had in the military service, but I would hate to have to go through it again." And then, after thinking a moment, he added, "that is, all except the leave time I spent in Paris."

After a prolonged illness, his adorable wife Edith passed away on Mother's Day, May 13th, 1979.

In 1980, he sold his business.

John lived until December 6, 1994. He was honored with a 21-gun Salute. Edith's son, Benjamin Christian Zinkhan III, passed away on March 2020. They are survived by one child, Julie Annette Mullenax VanMeter, and her son, John Dustin Eye.

OTHER SOURCES

Keir, Samuel M., Two Centuries of Valor: The Story of the 5th Infantry Regiment, Ch. 11

Wolf, Perry S., Fortune Favored the Brave; A History of the 334th Infantry 84th Division

I am now back across the 38th Parallel Line   Julie Mullenax Van Meter

The John Mullenax Family

Season's Greetings

I am now back across the 38th Parallel Line　　　　Julie Mullenax Van Meter

For you, my Angel Dear

I do believe that God above
Created you for me to love
He picked you out of all the rest
Because he knows I love you best
I once had a heart of mine, so true
But now it has gone from me to you.
Take good care of it, as I have done
For you have two, and I have none.
If I go to heaven, and you're not there
I'll paint your face on the golden statue
So, all the angels can know and see
Just how much you mean to me.
If you'd not come by Judgment Day
I'll know you have gone the other way.
So I'll give these angels back these wings,
Their golden halo's and other things
And just to show you what I'd do
I'd go to "Hell" dear just for you.

(From letter 137)